Edinburgh
Information Technology Series

COMPUTER DISCIPLINE AND

DESIGN PRACTICE

S. Michaelson, M. Steedman and Y. Wilks
Series Editors

AART BIJL

COMPUTER DISCIPLINE AND DESIGN PRACTICE

SHAPING OUR FUTURE

EDINBURGH UNIVERSITY PRESS

© Aart Bijl 1989
Edinburgh University Press
22 George Square, Edinburgh

Set in Lintronic Times by
Edinburgh University Press and
printed in Great Britain by
Redwood Burn Ltd, Trowbridge

British Library Cataloguing
  in Publication Data
Bijl, Aart
Computer discipline and design practice:
  shaping our future – (Edits)
1. Design. Applications of computer systems
I. Title II. Series
745.4′028′5
ISBN 0 85224 644 7

CONTENTS

	Preface	vii
1.	Computers in Design	1
	Design in mind	3
	Knowledge	6
	Design practice	7
	Advances in technology	11
	My experience as a designer	13
	Summary	21
2.	Language, Logic, and Intuition	23
	Design	24
	Language and intelligence	27
	Parts of language	28
	Definition and consequences	30
	Deeper into logic and expressions	36
	Knowledge	40
	Open and closed worlds	45
	Summary	49
3.	Design and Theory	53
	Design of buildings	54
	Other design fields	64
	Towards a theory	65
	Ambitions for CAD	69
	Quality	72
	Summary	79
4.	Integrated Design Systems	81
	Integrated systems	82
	Example of a user organisation	94
	Experience of users	98
	Software technology	107
	Lessons from experience	112
	Summary	114
5.	Function-Orientated Systems	116
	Task-specific systems	117
	Example of ground modelling	119
	Functions and intentions	128
	Why we have to bother	132
	Summary	134

6.	Drawing Systems	136
	Dumb systems	137
	Ordinary drawing practice	142
	Computerised drawing systems	154
	Summary	170
7.	Designing in Words and Pictures	172
	Prescriptiveness	173
	Descriptions	175
	Graphical concepts	182
	A worked example	190
	Summary	208
8.	Our Future	211
	Choices	212
	Technology of words	216
	Computer literacy	223
	Our future	226
	Summary	230
	Bibliography	231
	Index	235

PREFACE

I first thought of this book a long time ago, when I found myself between two contrary cultures. One was presented by people who are immersed in the science of computing, and the other by people who are meant simply to use computers. Added to that, there are the many people in between who live off the mismatch between these cultures: politicians, administrators, and entrepreneurs in the IT industry. From all this we get a tower of Babel producing noise that remains unintelligible to many participants. Keeping a research unit in funds has placed me in the midst of this noise. I felt that it should be possible to develop a more widely intelligible discussion of the conjunction of computers and people.

This book has undergone several transformations over the years. During that time I have drawn freely on experience of working with all the people who have been or still are members of EdCAAD at Edinburgh University. To name a few: David Rosenthal encouraged the view that we can re-shape computer technology; Fernando Pereira inspired us with confidence in our ability to contribute to the field of artificial intelligence; Sam Steel brought us his logican's mind; Ramesh Krishnamurti tried to add mathematical rigour to our theories; John Lee posed challenging questions from philosophy; Peter Szalapaj advanced the possibility of mixing logic and graphics; and Chris Tweed maintained our sense of relevance to design practice. I thank them all for our rich discussions which have influenced my writing, and I absolve them from any responsibility for what I have written. I also give special thanks to Margaret McDougall for keeping us in line, and for typing many versions of this text.

Thanks are also due to the sponsors who have supported our work, including the Scottish Special Housing Association, the Science and Engineering Research Council, and the European ESPRIT programme.

Lastly, much of this book has been written in what is strangely called spare time, meaning that time which rightly belongs to my family. For her understanding and patience, I give warm thanks to my wife, Valerie.

CHAPTER ONE

COMPUTERS IN DESIGN

They tell me that computers can know and do things, to help me design things: can we question the plausibility of this position? and where will our questions lead us?

We have now experienced over two decades of computer aided design (CAD) — ever since it was shown that computers can be made to draw.[1] In the late 1960s and early 1970s it was widely anticipated that CAD would revolutionise design practice and it was expected that within a decade or so computers would be used in most design offices. Over the intervening years CAD has latched on to new technological advances that have come from other fields of research, such as new computer programming languages, relational database systems and, more recently, logic programming and intelligent knowledge-based systems. Coupled to these software advances, there has been a steady increase in power and a reduction in cost of computer hardware, so that we now have microcomputers that match the power of the large computers of ten years ago.

Despite our expectations and our exploitation of these technological advances, CAD has still not fulfilled its promise. CAD is still not widely used to aid design, and this is especially true in loosely constrained fields such as building design. On certain aspects of buildings, such as their structural designs, computers are used to perform analytical tasks and their graphical abilities are used to produce visual presentations of design proposals. But computers are generally still not used to assist in generating design proposals, in synthesising designs.

What is it that limits progress in CAD — its practical application in design offices? Do we have to accept a partial fulfilment of the promise of CAD and accept the emphasis that places on analytical techniques applied to design? Can we be satisfied by demarcations between the roles of analysis and synthesis to differentiate between the roles of computers and designers? Will such a distinction lead to changes in the role of design as perceived by people in general, and will people want that?

Is the future of CAD dependent on technological advances? Do we have to wait for technical solutions to computational problems to obtain computers that will satisfy our expectations for CAD? Should we, instead, recognise that the ambition of CAD extends beyond technology and requires us to reassess our understanding of design? Should we be looking more at people, looking upon designers as 'machines' for designing, to inform our reasonable expectations for computers?

These questions have prompted this book. The following chapters are based on my experience as a practising designer, followed by twenty years of research into computer aided design. My research experience has progressed from ambitious computer applications towards an increasingly theoretical focus on artificial intelligence (AI) and cognitive science directed at natural language. Early efforts on CAD applications prompted me to relate the broad concerns of practising designers to the separately focused disciplines of other academics, to their various approaches to knowledge. Experience of applications led firstly to my search for deeper understanding of the formal approaches employed by philosophers, logicians, and researchers in AI. I have found myself questioning the relationship between formalisms and ordinary activities of practical people, leading to a fundamental reappraisal of the roles of knowledge represented within computers and knowledge existing within people.

In presenting my conclusions, I am faced with a language problem; language relies on shared perceptions among people and I am having to refer to different perceptions of separate groups of people. Academics in separate disciplines do not find it easy to talk to each other and, collectively, they are not good at talking to 'ordinary' practical people. Typically, academics expect references to substantial bodies of literature which support their own disciplines; practical people prefer more informal references to personal experience. I will attempt to meet these expectations by referring to my own experience and the experience of users of CAD, linked to my understanding of formal disciplines. My broad spread of concerns may lack academic authority; some will argue that I should have included more references to the writings of others and, no doubt, my presentation develops ideas that have come from other people whom I have not referenced. To counter such criticisms, I can only reiterate that I am searching for a language that can be understood by dissimilar groups of people.

Computers in Design

My intention in writing this book is to build a bridge between the formal knowledge of academics engaged in new technological advances and the world of all other differently specialised people, the world in which technology is employed. The more formal reader can expect to find clues that will prompt new efforts in research, and may find some of my arguments contentious. The general reader, and especially designers, can expect to discover insights into the concepts that underlie CAD and how these concepts condition the usability of systems that people are invited to use. My reason for trying to bridge different kinds of knowledge is that future CAD and computer technology in general will affect all our lives in ways which we will or will not like. I believe that everyone has an interest in foreseeing and shaping that future.

DESIGN IN MIND

Design refers to a human activity which takes place in the minds of people, and which we do not know how to externalise as processes executed outside people. My focus is on synthesis of design objects, any objects, as in the architecture of buildings. This reference to an activity of the mind indicates a focus on the individuality of persons, which leads us to consider the discipline imposed by computers, the mode of thought they require of their users, and the role of knowledge detached from people. These issues gain importance when we observe design practices in the world of people, and they are highlighted by the example of design in buildings.

The issues to emerge can be viewed as general and familiar issues which have been aired before, often in the course of heated debates but without permanent conclusions. Examples are found in the variously focused debates on man versus machine and, more recently, on the feasibility of intelligent machines versus intelligence as a unique attribute of people.[2] The intention in the following chapters is not to take sides or to prove that any one position expressed by these polarities is objectively correct or morally justified. Instead, the intention is to steer a course between them, to consider how different philosophical beliefs have practical outcomes that condition ordinary human activities. These beliefs can be shown to affect advances in computer technology; they present us with choices and we have to decide what we want to do.

All the discussions developed in this book are founded on the following initial positions, with regard to philosophy, artificial intelligence, and design practice.

Philosophy

Philosophers seek reasoned explanations for all human experience of our world to establish what we can know about this world. This ambition embraces knowledge itself, different categories of knowledge, and the status that people attribute to various kinds of knowledge. It also embraces assumptions or beliefs on which knowledge is constructed, and the motivations for adhering to selected beliefs.

My position with regard to belief is essentially agnostic. We cannot know that particular beliefs are correct in any absolute sense by reference to things inside or outside our world, but we can judge the effects of beliefs in terms of what they enable us to do. Agnosticism here refers to the possibility that any current beliefs may be wrong, and we judge them to be wrong when we do not like the things they make us do. We should be willing to question any beliefs and explore their consequences.

This position may not warrant the elevated title of a philosophy; it is motivated by interests in practical outcomes. The following discussions are intended to emphasise direct links between philosophical stances and the practical things we do. The conclusion to emerge is not that we need to adopt a particular new philosophy, but that we should look upon all philosophies as reflecting certain states of human understanding that condition human activities.

Artificial intelligence

Our general ambition for computers rests on a broad and popular belief that computers can be made to exhibit knowledge. Subject to various qualifications, this belief is necessary for all applications of computers. The field of AI adopts a relatively extreme position, focusing on computational approaches to intelligence with the ambition of making computers behave in ways that people will recognise as being intelligent.

In all cases, efforts to make computers do things rest on our abilities to externalise knowledge outside people. These efforts pose questions about the independent existence of such externalised knowledge, its subsequent autonomous evolution, and the responsibility and authority which we can attach to such knowledge. These developments can be viewed as following in a long human tradition of striving for objective knowledge.

By questioning the assumptions that support the ambition of intelligent computers, we will find ourselves questioning established beliefs. Arguments may appear unorthodox and are likely to attract the kind of ridicule that sometimes accompanies religious controversies. Again, my position with regard to the possibility of intelligent computers is essentially agnostic. We need to know what we mean by intelligence in order to explore possible demarcations between human and artificial intelligence.

Design practice
Design is an activity which lies somewhere between calculated reason and magic, calling on individual judgement and responsiveness from within persons. An important task of designers is to give expression to values and aspirations of other people. At best, a design should be a tangible celebration of being alive.

By calling on judgement, beyond calculation, design is similar to all ordinary human activities; it differs only in the extent to which it is expected to evoke notions within people. This similarity with ordinary activities means that any issues to emerge from a conjunction of designers and computers are likely to have general impact on the future of computers and all people. The necessary individuality of designers poses questions about the role of overt and shared knowledge, and any partitioning of intelligence between computers and people.

By focusing on the example of design practices targeted at buildings, I will highlight these issues. However, my answers are not intended to be conclusive. For purposes of realising artificial intelligence, we can create possibilities by deciding upon definitions for intelligence that circumscribe and thereby limit our appreciation of abilities within people, and we then have to decide whether we want those possibilities. Alternatively, the valued practical contribution from the field of AI may eventually prove to be a richer appreciation of people's intelligent use of dumb machines.

By way of introduction, the whole of the argument presented in the following chapters will be summarised in the next paragraphs. I will focus on the enabling power of mental mechanics, referring to the overt and detached nature of formal mechanisms which can be used to process information. I will consider their implications on the involved nature of people-centred responsibility for decisions and actions among people. Underlying this theme is an acceptance that considerations of quality and human welfare are of paramount importance. These considerations depend upon holistic contributions by people — individuals matter. Advances in knowledge and associated technological developments should not be divorced from the people who promote them. Accordingly, this chapter closes with a brief sketch of myself as a person, a designer, and a computer technologist.

KNOWLEDGE

Chapter 2 considers how we know things and how knowledge is represented. It develops the view that human knowledge is part of being human and that externalised representations of knowledge are not the same as human knowledge. Representations, like models, are partial abstractions and we cannot be certain about the demarcation between knowledge and any part which might be included in a representation. Representations and devices for operating on representations, as in the case of computers, can be useful to people if they satisfy knowledge within persons and that qualification has to remain subject to human judgement.

This is a philosophical position which admits the possibility of other non-human knowledge as part of beings that are not human. However, dissimilarity between beings is likely to reduce their recognition of each other's knowledge — there might be such a thing as autonomous artificial intelligence, but we are unlikely to know what that can be. In this chapter I advocate people's intelligent use of dumb machines.

Computing and design both refer to kinds of mental activity, but with important differences. Computers are part of a long tradition of people's endeavour to externalise thoughts, so that thoughts can be passed between persons in the form of overt knowledge. This tradition embraces the objectivity of natural sciences, the detachment of mathematics, and the truths of formal logic. It includes the kind of knowledge that can be represented by means of abstract symbolic expressions, such as words and numbers. It has now come to be accepted as setting the ground rules for our rationalisations and our justifications for all things we can *know* and *do*. Computers are the latest development in this tradition, and they are advocated as devices in which knowledge can be represented independently of people, which can operate upon knowledge autonomously on behalf, or in place, of people.

Design, on the other hand, differs in the way in which it employs people's intuitive knowledge, embracing all kinds of human experience. It calls on holistic involvement of persons beyond their use of formal, overt representations of knowledge. Design relies on human judgement to tell us what we *want* to do. This difference can be emphasised by adding that design activity calls on human sensitivities, on powers of perception and interpretation that are not bounded by conventions governing the formal correctness of symbolic expressions. Design tends to employ analogic forms of depiction, involving interpretation of people's concerns directly into artefacts. Correctness of results or, more precisely, the 'goodness' of designs is always subject to human judgement, never proved.

The difference between *overt* and *intuitive knowledge* is fundamentally important to any view of design, and any other human activity. We have to try to be clear about how we use this distinction when we devise artefacts to operate on overt expressions of knowledge. Systems for constructing and manipulating expressions can be regarded as mental mechanics, enabling us to express what we think overtly. Like physical mechanisms, such mental mechanics condition how we can do things, and what we can do. In neither the physical nor the mental sense do the mechanics themselves justify what we actually do. That actuality is governed by human considerations outside the mechanics, which prompt our development of new mechanisms. However, an appreciation of these systems does have general importance, affecting our understanding of our world, our place within this world, and what we do with it.[3] More specifically, these issues bear on design and on CAD. We have to answer the crucial question: what interaction can exist between formal systems operating within computers and the informal behaviour of designers who are invited to use computers?

DESIGN PRACTICE

Chapter 3 develops a concept of design as applied to the architecture of buildings, and observations are generalised towards a theory of design. Design is regarded as something people do — the task of conceiving and expressing new models that cannot otherwise be deduced from pre-existing models, which can apply to both artistic and scientific endeavours. The finished products of designing, designed artefacts, on their own do not tell us much about how artefacts are designed. Instead, we have to consider the world of design as being the thoughts in the minds of designers, the receptiveness of those thoughts to the needs and aspirations of other people, and the ability of designers to externalise their thoughts. CAD ought to offer techniques by which designers can exercise and externalise any thoughts.

The act of designing can be characterised, as a process of *synthesis*. Design has to generate a fusion between things to create new things, so that the products are recognised as having a right and proper place in a world of people. Things should be understood as referring to anything — physical objects, abstract ideas, aspirations. These things are extracted from some design context, transformed through designing, and results replaced. In the example of buildings, the context is people and results have to be assessed by reference to people.

This generative view of design differs sharply from the more orthodox *analytical* approach of design systems. In the latter case, the usual strategy is to decompose a given problem into parts until separate parts are recognised as being amenable to known evaluative procedures, and results are then aggregated into a solution. This approach has a peripheral role in design, when evaluating selected aspects of tentative design proposals. However, the absence of overt and widely recognised criteria for solutions excludes design from the mainstream of analytical developments.

The following chapters consider past experiences of three kinds of computer application: integrated design systems, function-orientated systems, and drawing systems. This experience forms the basis for more general concerns expressed about computer technology, focusing on possible interactions between users' perceptions of their world and representations of such perceptions within computers.

Integrated design systems
Chapter 4 discusses *integrated design systems*. This kind of system employs a single model that can accommodate all information describing a design object and the model corresponds to knowledge supplied by different people. The model then has to be capable of supporting a range of operations on the same description to advance people's different interests during the course of designing the object. In addition, the system has to be capable of interpreting information from people's overt expressions of their thoughts about a design object, from text and drawings, to supply that information to the model. This ambition focuses attention on the possible ways of organising data within computers, a data-orientated approach.

Despite the success of early examples of these systems in practical use, they did not mark the beginning of a new era in CAD. The most serious problem they presented, and which persists to the present day, is that they relied on a very close correspondence between a user's design practices and the modelling procedures built into the system. Those procedures had to be defined at the time when each system was being developed. Predefined computer models had to anticipate precisely how buildings would be perceived in the users' world, down to a level of fine detail. Users had then to conform to these anticipations, leading to these systems being characterised as *prescriptive*. Worries about prescriptiveness were expressed by the authors of one of these systems in the mid-1970s:[4]

In general, our view of design practices suggests that design is not a knowledge based discipline, but evolves from the experience of many individual practitioners. By this we mean that the knowledge used by designers cannot, on its own, be constituted as a formal model which will then be recognised by all designers as representing their own individual practices. The role of experience is to build up the responses of designers to the tasks which other people present to them, and experience results in unique combinations of formal knowledge and subjective understanding within individual designers. This conclusion has profound implications for future, generalised CAD systems.

Function-orientated systems
Chapter 5 discusses *function-orientated systems*. This kind of system relies on an anticipation of some specific task in the users' world as the primary motivation for its development. An example might be the calculations required for environmental appraisals to assess performance of proposed designs for buildings. Anticipation of task is paramount, providing a specification of required system operations, and organisation of data is subservient to the task. Typically, these systems perform discrete tasks, each requiring separate input and providing output which users have to translate into their own perceptions of design.

Systems were developed which incorporated sophisticated simulations and analytical tasks, so that computers increasingly came to be regarded as doing the work of consultants, evaluating specific aspects of building designs. Problems were presented by these systems, by the boundaries to discrete tasks and associated concepts of classification and typing which had to correspond to perceptions of different designers. As a general and persistent example, the boundaries between quantitative and qualitative evaluations are notoriously variable for different instances of design. A still more serious problem, referring to the separation of knowledge within computers, has been expressed as:[5]

> the idea that you can encapsulate a specialist's knowledge inside a computer program, such that an architect can approach the program and not speak to the specialist. Typically, when an architect consults a specialist he is seeking access to knowledge possessed by that other person. If the architect is looking for an environmental evaluation, to get guidance on that aspect of a design, the specialist will come up with answers and they may appear as numbers, resulting from a disciplined

view of environment. The architect then has to make those answers fit in with all other considerations that bear on a building design. The building is not only going to perform in terms of environmental behaviour, it has to do many other things. The answer may not fit. What then happens in conventional practice is that the architect starts arguing with the specialist. What he says is — 'that answer might be right, but it does not fit, it is not good enough . . . how far can I go wrong to find an answer that will fit in with all the other considerations that have to be expressed in this design?' What you are doing is asking the specialist to break his rules, to see how far you can change his answer. The specialist has then to step outside the bounds of his discipline and refer to his individual experience. This bartering process that occurs between architect and specialist cannot, as yet, be encapsulated in a computer program and the specialist needs to be a human being.

Drawing systems
Chapter 6 discusses *drawing systems*. Following on the experience of integrated design systems during the 1970s and the limited success of function-orientated systems, new CAD developments of the late 1970s and 1980s became far less ambitious. This trend was marked by the emergence and the popularisation of drawing systems. These systems deal with drawings in much the same way as word-processors process words. The key difference to earlier integrated design systems is that, in the case of drawing systems, the computer knows nothing (or very little) about what is being depicted in drawings; the computer knows only about lines and points, and about edits and transformations that can be applied to collections of lines. The aim of these systems is to produce drawings faster and of better quality than might otherwise be produced by hand.

Like word-processors, drawing systems are potentially useful to many designers. As may be true for any generalised system, this potential depends on drawing systems remaining dumb, so as not to conflict with different knowledge possessed by designers. Past systems that have been powerful enough to support production of full working drawings for all kinds of buildings have presented the dual problem of high acquisition cost and high investment in learning to use them to draw in accordance with particular design practices. High investment has resulted in systems being set up in offices that offer drawing services to other design offices — posing questions about how those other offices can then express their own design knowledge.

ADVANCES IN TECHNOLOGY
Experience of CAD points to inadequacies of *prescriptive* computer technology. Prescriptiveness presents problems whenever it results in computer programs that encapsulate users' domain knowledge, meaning users' knowledge of application fields. Prescriptive programs inhibit the evolution of people's knowledge. We can then observe that there is a fundamental incompatibility between the irregularity of demands that people make on designers, and the separate expectation that designers' work-practices should conform to predefined procedures set up within computer systems. Prescribed forms of expression which are defined by the need to access predefined organisations of design knowledge held within computers are liable to conflict with the ways in which architects have to think and express their thoughts, when designing.

These issues cannot be resolved simply by writing better computer programs. Even the change in technology that is marked by the shift from deterministic procedures to more open-ended rule-based systems is not likely to prove sufficient. This latter development is evident in expert systems and more generally in the field of AI.[6]

Description systems
Chapter 7 develops a possible general strategy for CAD. This rests on an understanding of *descriptions* as expressions that can describe anything. Descriptions are regarded as a kind of expression which is free of any external criteria for correctness, is temporal and arbitrarily modifiable, and which remains the responsibility of its author. Descriptions, as in the form of text and drawings, are considered as compositions of objects and relations, such as arrangements of characters and lines. Definitions of those constituents are conditioned by the actions that are executable by authors and recipients, by people and by computers. Use of these constituents is conditioned by interpretations within people, and by received mappings within computers.

When a computer is used to produce descriptions, the basic constituents of the description environment and the operations that can be applied to descriptions are necessarily prescriptive. What distinguishes a computerised description system is that its prescriptiveness is targeted at construction and modification of descriptions, and not at interpretation of descriptions into people's domain knowledge. Construction here does not include the act of deciding what any particular composition of objects must be, nor does it include reading what a composition might mean outside the description environment. The aim is to achieve computer systems by which people can shape their own descriptions, thereby exercising and exhibiting their own intelligence to other people. Systems used in this way can then be made to reflect unexplained human intelligence.

This chapter discusses how graphical concepts can be realised in drawings, resulting in a general structure for arbitrary line drawings which can be represented logically within computers. The same logical representation scheme can also be used to represent arbitrary arrangements of text. This development opens up the possibility of producing integrated graphical and textual descriptions composed from logical relationships between character strings and assemblies of lines. Users can specify any properties of, and relationships between, any parts of text and parts of drawings to produce any unrehearsed descriptions.

The strategy for CAD which emerges here is motivated by the need for designers to be able to externalise any thoughts, to generate new design objects in response to unforeseen demands from other people. This strategy can be summarised in terms of the following general understanding:

(*a*) descriptions, expressed in some externalised form, can be defined in terms of some structure of objects and relations which are amenable to certain actions;

(*b*) these objects, relations, and actions can be represented in computational logic, in a computer, and they can be interpreted by people as representing human knowledge;

(*c*) a computer, used in this way, knows only of objects, relations, and actions that constitute a description environment, and it knows nothing of people's interpretations for descriptions;

(*d*) people can then use the computer to produce and modify descriptions of their own knowledge, so that these descriptions can be interpreted by other people into their further knowledge;

(*e*) to use a computer in this way, people have to differentiate between domain context-free concepts that condition descriptions, and domain context-sensitive concepts that condition interpretations — if we can do so, then we can use computers to describe anything.

A description system has to provide an environment in which designers can construct and evolve descriptions. A computerised description system needs to employ logic in a manner which remains visible and accessible to users, to allow them to formulate *their* expressions of design knowledge, and enable them to call on *their* machine representations of that knowledge in the course of progressing expressions towards design products. The implications for what users have to know in order to use a system in this way are formidable.

Wider implications
Chapter 8 considers the wider implications of description systems and computer technology generally. The effort of learning to work within a logic system may prove to be an unavoidable cost of people's progress towards computer literacy. We have the parallel case of word literacy and the evolution of people's ability to work within written word systems — the technology of words. I conclude this chapter with some speculations on the further implications of computer literacy on the evolution of human knowledge, politics and power; and I indicate some of the opportunities it promises for designers.

My Experience as a Designer

Advances in knowledge and associated technological development should not be divorced from the individuals who promote them. Accordingly, here follows a brief sketch of myself as a person, a designer, and a computer technologist. The general reader may be reassured by the fact that I, as the author of this book, am not a formally trained technologist — although my long association with such people may be more apparent in my writing than I would wish.

Born in the Netherlands, I started a fairly ordinary upbringing. Then the wartime years of my childhood were spent in South Africa, troubled with illness and nurtured on the tensions of a socially unjust regime, while enjoying a magnificent country and benefiting from the privilege afforded to my white skin. During my adolescence in London, I thrived on those post-war years of austerity. Back to Africa, at the University of Cape Town I became an architect. I found myself being transformed into a specialist — in those days we believed in the special vocation of architects to shape the lives of all other people.

I became what is sometimes called a designer's architect; one who is valued for intuitive design flare. I was impatient to design actual buildings and, before graduating, set up the office of Bijl, Green & Todeschini, with two other students. Examples of the products of this office, from 1959 to 1961, are illustrated in Figures 1.1 to 1.7. My final student project (Figure 1.1), a nightclub at the rock line half way up Table Mountain, was a bit of fantasy which successfully skirted around the established rules of design then current at the school. The coffee bar (Figure 1.2) was our first commissioned project in the middle of Cape Town's theatre land; it was judged to be unbuildable so we set up a workshop and built it ourselves, nights. Having our own workshop, we continued to sculpt in welded scrap metal (Figure 1.3). The church (Figure 1.4) was warmly received by the Catholic clergy for whom it was designed, but did not materialise due to ideological conflict with the man who had the money — an early lesson in the cost of principles, and their importance to self-preservation. Our main work was concentrated on housing for individual clients (Figure 1.6), and for developers (Figure 1.7). The most prominent example was for a timber millionaire (Figure 1.5), an exhibition of timber in the form of a hyperbolic paraboloid roof which, unknown to us at the time, gained fame as the largest in the southern hemisphere. Our office was closed by Sharpville — for those too young to remember, Sharpville was the event when government police shot blacks in the back, causing a ripple in a pattern of violence that momentarily halted investment in building. In retrospect, this event marked the end of my design career.

These examples are not intended to prove that I was a good designer, the reader may well come to a contrary judgement. They are intended to show only that I was a designer familiar with issues that occupy designers, and this knowledge has stayed with me throughout my subsequent career in research. This background is unusual for a researcher engaged in CAD and it has influenced my particular approach in bridging academic concerns and the interests of practising designers.

As a designer wishing to defend the sensitivities that I believed were important to myself and other people, I remember being resistant to what I saw as impersonal technological change. As a student, I strenuously defended the ruling pen against the advance of the Graphos. Later, I did succumb to the Radiograph — these were all bits of technology aimed at putting ink onto paper. By the time we set up our office, I had also reluctantly made the transition from T-squares and drawing boards to elaborate mechanical drawing machines. This reluctance to accept technological advance rested on a wish to preserve idiosyncratic control of expression, to permit individual design concerns to find expression in drawings. The attraction of technology was the power it afforded in enabling me to process more information. I was able to produce more of my own drawings.

Scepticism about technological advances has remained with me,

Computers in Design

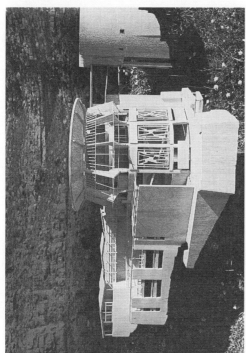

Figure 1.1 *A nightclub on Table Mountain*. A fantasy chosen to avoid the conventional tests for good design that were applied to sensible buildings — the author's graduate thesis project, 1959.

Figure 1.2 *LA Perla in Cape Town's theatreland*. Our first commission and no one would build it, so we used our hands to make the metal web ceiling, the mosaic on the counter and column, the vitreous table tops, ash trays, metal table lamps, and wall tat — this and the following figures illustrate work of the office of Bijl, Green & Todeschini, 1957–61.

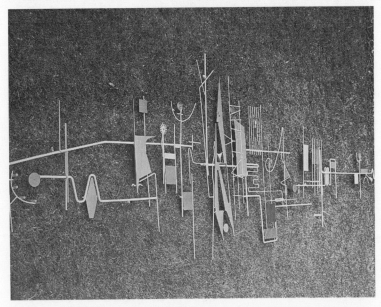

Figure 1.3 *Metal wall sculpture for v Niekerk, Karoo*. The reputation of La Perla kept our hands busy making things as well as drawing.

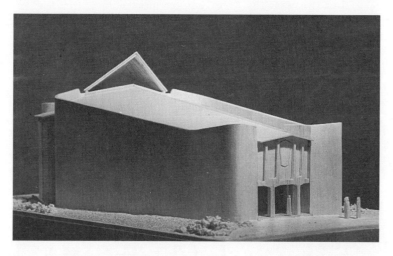

Figure 1.4 *Catholic church, Newlands*. The expressiveness of this design was appreciated by the clergy, but we disagreed with the financier over his wish for a Spanish villa, so both were not built — a lesson in principles and self-preservation.

as is evident in my critical approach to CAD research and its achievements. But please do not confuse criticism with condemnation. I see criticism as essential to research, opening up questions that need to be explored, often leading to more questions. Searching criticism is the best protection we have to ensure that necessary advances are also compatible with people.

Transition from practice to research
After settling in the UK, my transition to academic research was prompted by a wish to discover more about the nature of ideas that link human concerns and design practices, in the face of popular enthusiasm for design methodologies in the 1960s. I could not reconcile what was then called rational methods with my observations of design practices. My first exercise in research was an exploration of all decisions that affected a large urban redevelopment project, in which I sought to identify some coherent order in the contributions from all participants, and failed.[7] Computers offered a means for conducting further explorations, promising rigorous tests of the rationality of systems, and providing opportunities for studying the effects of rationalised systems on human activities.

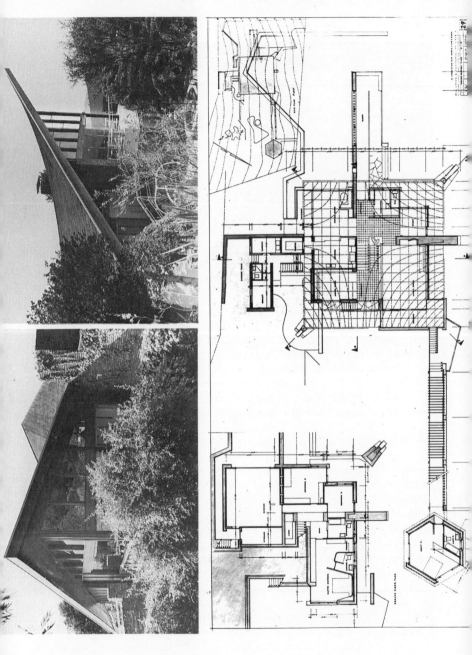

Computers in Design

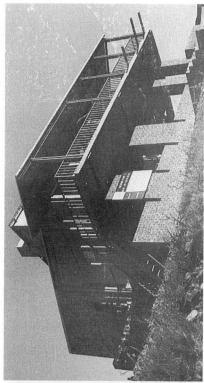

Figure 1.5 *House Bruynzeel in Stellenbosh*. An exuberance for an international timber merchant — straight lines of wood in the curved hyperbolic paraboloid roof (above), picked out in the straight lines of the plan (below), producing dramatic but simple shapes to the spaces inside the building, with curved yellow-wood ceilings sweeping up and outwards.

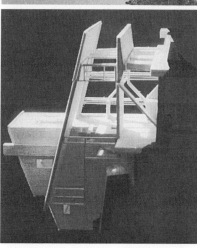

Figure 1.6 *House, Ribiero, Llandudno*. An example of a house built for a private client, the kind of project that used to be abundant for young designers to grow on, to develop their design skills.

Figure 1.7 *Balgay Court, Kenilworth*. A project for a developer, with some scope to explore design sensitivities; the kind of project intended to provide bread money but, in our case, this intention collapsed at the time of Sharpville.

On entering the field of CAD in 1967, with the courage of ignorance I set about developing computer applications, targeting available technology at ambitious systems, and I rapidly gained a sense of achievement. Then, experience of systems in use gave rise to criticism and dissatisfaction with the underlying technology. I plunged deeper into what might loosely be called computer science issues. Finally, I found myself forced to examine the concepts underlying computer technology and relate these to my concept of design. I had to recognise the crucial role of theory. This progression from applications to more fundamental research followed a familiar path among CAD developers, and presented the classic difficulty of knowing where to stop. In my case, I chose to probe fundamental questions as far as my mind would allow.

Over the past twenty years, I have tried to avoid becoming a specialist in any one of the many fields that can contribute to the goal of CAD. Specialisms are necessarily exclusive, relying on some narrowly defined target that cannot embrace all that is explicit and implicit in a broad goal. However, I believe that all the ideas underlying each specialism can be expressed and understood in terms of common sense sufficiently so as to relate the ideas and assess their implications for the broad goal. This is an ambitious belief which entails hard and painful effort. One has to avoid the temptation to identify ideas simply in order to categorise and reject them. The aim is to keep ideas alive and make constructive connections between them, while fending off parochial criticisms from people in those fields where the ideas have originated.

I find computers as artefacts uninteresting, often irritating. Computers as implementations of ideas are interesting, and we can explore these ideas without knowing much about the physical devices. My intention is to raise these ideas up to a level of common sense, so that ordinary people can judge what computers should be able to do.

Summary

This introduction has outlined the main theme of the book. We have now experienced two decades of CAD. In that period, the promise of CAD has met with obstacles presented by, first, the essentially idiosyncratic nature of design practices, particularly in loosely constrained fields; secondly, the prescriptive use of current computer technology; and, more recently, the assumptions underlying machine intelligence. It is time to review our fundamental assumptions for CAD.

Design focuses attention on the point of contact between

detached scientific knowledge and people's ordinary concerns — a complex ordinariness that embraces subjective motivations and aesthetic sensitivities. The issues that are presented by people being answerable to other people are discernible in examples of past integrated design systems, function-orientated systems, and drawing systems. The boundedness of overt knowledge, as encapsulated in computer systems, then presents problems when designers have to respond to unforeseen demands from other people.

The orthodox assumption that knowledge is objective lends credibility to the authority that is commonly associated with overt knowledge. The same assumption also supports the ambition that people's domain knowledge should be represented and exercised autonomously within computers. It is this orthodox assumption and its effects which need to be re-examined. It is possible to conceive an alternative assumption. By differentiating between human knowledge, overt expressions that emanate from such knowledge, and overt logic systems that support manipulations of such expressions we can then conceive that human knowledge exists only within human beings — not in expressions or in logic systems. Logic can then be viewed as providing mechanisms for manipulating expressions, without responsibility for correctness and untroubled by ambiguity in the world of people.

Notes to Chapter 1

1. Early impetus for CAD came from work by Sutherland (1963) and others at MIT, on Sketchpad.
2. Combatants in the general arena of man versus machine have included people such as Bronowski (1971) and Schumacher (1974) and, in the more focused debate on artificial intelligence, include people such as Dreyfus (1979) and Searle (1984). For an easily readable introduction to the various deeper philosophical stances which bear on these issues, see Magee (1987).
3. Here I refer to the kind of questions posed by Schumacher (1974) on technological know-how, what he calls bad metaphysics, and the need to know-what. For a designer's view of similar considerations of quality, see Pye (1978).
4. Report for the Department of the Environment (Bijl *et al.*, 1979).
5. Paper presented at a meeting of the Royal Institute of Swedish Architects (Bijl, 1983).
6. For an early and far-sighted discussion of the application of AI to architecture, see Negroponte (1970).
7. Report for the then Ministry of Public Building and Works (Bijl, 1968).

CHAPTER TWO

LANGUAGE, LOGIC, AND INTUITION

Language enables communication between natural abilities in intelligent beings — how, then, can we communicate with computers?: we need to consider intelligence and language in the context of design, and the feasibility of recreating intelligence in computers.

In this chapter I consider how we *know* things, the extent to which such knowledge can be *externalised* outside ourselves, and our use of language to *express* knowledge. It is a personal statement of a philosophical position which is important to my presentation of experience in the following chapters. This position is not unique, but it runs counter to current popular belief in knowledge being founded on hard facts, which supports the ambition of knowledge being replicated within computer systems. Those readers who find these issues obscure and irritating should feel free to skip to the next chapter.[1]

We generally accept that there is a close link between knowledge and language. We all use language to reach into those abilities that exist within each other. We do so in order to condition our actions; so that we can recognise each other's actions as intelligent behaviour. Our language evidently works within and across varied discourse domains, including design. Can such use of language survive the formal treatments of language aimed at users of computers? Alternatively, are there boundaries to what we can know about language which will limit the role that can reasonably be attributed to computers?

Here we have some key issues of relevance to computer technology. Answers are being sought through philosophical investigations and by researchers in this field. Technical solutions are being developed for certain kinds of application and promises are being directed at an ever increasing range of applications. Popular advocacy of the technology seems to be running ahead of our understanding of possibilities. Design, then, provides an interesting and testing context in which to explore the key issues. Design points to a use of language which is particularly difficult to reflect within computers, and a use of language which is familiar in many people's ordinary activities. We want to know if computers necessarily have to impose a different use of language and, if so, we want to know what that difference will mean for intelligent human behaviour.

DESIGN

To start this discussion, I will introduce a view of design that comes from my experience of architecture, CAD, AI, and my recent work with linguists.[2] Essentially, the language of design as applied to the architecture of buildings rests on an old (and recently suppressed) distinction between buildings as artefacts and architecture as a property of these artefacts. Briefly, buildings can be regarded as aggregations of parts with their associated functional properties (the bits that keep the water out, the temperature in, for example). Architecture is read into the resultant whole compositions, in the spatial modelling of deep aesthetic sensitivities (embracing religious, social, and economic concerns, and including 'style' conditioned by 'market forces').

Architecture is an inescapable property of those things that we call buildings. The architecture of buildings is what people perceive when they assign value to buildings. All people do so. Architects are specialists in formulating architecture to serve as expressions that can be appreciated by other people. Architects who use the term architecture selectively by saying that certain buildings are not architecture imply a distinction between good and bad architecture. When other people read bad architecture into buildings, then buildings fail. Failure leads to neglect. Here we come to a general (and perhaps contentious) dictum: buildings might leak if and after they have failed architecturally, rather than fail because they leak.

This initial and brief characterisation of design as applied to the architecture of buildings points to the importance of design and emphasises the role of people in design (Chapter 3). Why people want buildings, what they expect buildings to be, and what they intend to achieve with buildings, change as people change. Architecture cannot be produced solely by means of prior, deductive, goal-driven techniques. Instead, architecture requires a mode of thought that can exploit individual experience and intuitive judgement, operating at the conjunction between detached scientific knowledge and involved human feeling. This is the mode of thought which is central to any act of designing.

Computer aided design
When we turn to computers, we need a slightly more abstract and general appreciation of design. We can approach this position by asking a general question: what distinguishes CAD from other uses of computers? I suggest that there are two distinctive features of CAD.

First, CAD applies in situations where information demands action but where there is no prior model for organising the information and determining the action. In this sense, design is creative. A design process, whether in a person or in a computer, has to allow the creation of new models to accommodate information in some orderly form. Designers' products are models. This is true for the design of buildings, where information defining demand may be disorderly and loosely bounded, and where the response has to be a model (the drawings and specification) for a building. This also is true for new collections of scientific data, where there is no pre-existing order for the data, and where the response has to be a model (an equation) that will make the data meaningful to some intended purpose. The motivation for a new model generally comes from outside the computer system, from the user of the system. A CAD system must be capable of receiving new arrangements of expressions that will result in new design models within the system.

Models, as design objects, are constructed in the system's knowledge base (Figure 2.1) and are different from models of expert knowledge (shown under 'expertise', Figure 2.2). The latter are prior models against which a proposed design object may be evaluated and tested. Certain design models may acquire the status of expert knowledge (as in the case of scientific models) if they are useful to the validation of further design objects. The point here is that design objects cannot be determined solely by models of expert knowledge — that would constitute problem solving and not design generation.

Following on this last point, building up a new model in a computer involves a user in programming the computer. The user has to specify the model in terms of lower level functionalities that are already available within the computer. A CAD system, therefore, has to provide functionalities that are recognisable to the user as being meaningful to the user's own interests — a modelling environment.

We now come to the second distinctive feature of CAD, which is

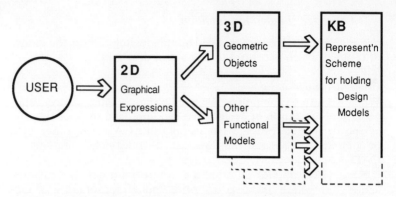

Figure 2.1 *Components of a CAD System*. Expressions are the tangible means by which a computer system and a user can communicate with each other. Interpretations from the user's expressions are passed to functional components of the system. These components represent expertise (Figure 2.2) on aspects of descriptions and performance of design objects. Further outcomes are held as design models in the system's knowledge base. The (reversible) bold arrows indicate a flow of interpretations between expressions and models.

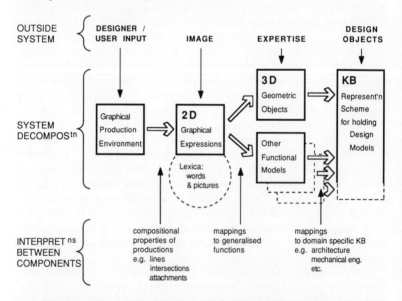

Figure 2.2 *Systems for processing expressions*. Here we have an expansion of Figure 2.1. Emphasis has been given to the production of expressions, by adding a component that serves as a graphical production machine. Other parts of the system, and people who use the system, work through this component to produce expressions which pass between the system and the user. The whole system can be viewed as a system for processing users' expressions which are intended to be passed between people.

Language, Logic, and Intuition

that expressions which pass between users and a computer system typically need to be in graphical form. Differentiations within depictions denote values for spatial properties and relations, and these differentiations ought to be significant to representations of spatial and other knowledge held within the system, that comes from users. This use of graphics places emphasis on graphics as language.

Design in both architecture and CAD places emphasis on the role of people, which causes us to focus our attention on how people are able to communicate with each other. We are prompted, therefore, to consider the naturalness of 'natural' language.

LANGUAGE AND INTELLIGENCE

The ability to use language is widely considered to be a distinctive ability of intelligent beings, which separates us from lesser animals. This link between language and intelligence supposes some functionality of language which, in turn, supposes some system by which language operates. We think of the system as being based on logic which serves to establish truth conditions for expressions. Language is then conceived as a means for conveying intelligence between intelligent beings.

This fairly orthodox view of language poses many questions. Does it rest on a humanly biased view of intelligent beings, embracing only those beings that are most like us and think like us? Can we think of other forms of intelligence that are unintelligible to us? If we cannot does that preclude the possibility that such other intelligence exists? These general questions become more pertinent when we apply them to a world of dissimilar people. When people with differently developed minds, such as formal theoreticians and intuitive designers, are obliged to communicate, they might experience difficulty in recognising each other's intelligence. Yet, in order to proceed, they have to accept the probability that each is intelligent, otherwise they have to disengage, or resort to some more direct physical interaction.

We usually acknowledge the presence of diverse intelligence in people by saying that people exhibit variously specialised uses of language. Is there some common basis to the functionality of language, which binds together all forms of intelligence, or at least human intelligence? Is logic that basis? What does logic have to be if it serves this purpose?

Broadly, logic can be considered as offering reasoning mechanisms for constructing representations of human understanding. We use logic to establish what can be and, set in the context of time, what has been and what will be. Thus, we have the topical 'true if' and 'if true then' deductive mechanisms, with refinements to differentiate between 'true for all' and 'true for some', or 'true for now' and 'possibly true later'. Can these be universal mechanisms and can we describe them in some expanded form to encompass all human intelligence. Furthermore, are these mechanisms inherently conditioned by the kind of thing we are, and can we expect to establish that fact?

What is the relation between logic and expressions where expressions are the tangible manifestation of language? Commonly, it is claimed that there is a strong connection between logic and those expressions that are in the form of words, either spoken or written — concepts that are amenable to logical treatment are expressed in words (or in some more formal symbolic or mathematical notation). Does this have to be an exclusive connection? Can different logics be defined on different forms of expression, such as pictures? Can there be mappings across different forms of expression, for purposes of representing composite concepts and invoking complex logic? Evidently people use pictures to express intelligence.

What are the criteria for the successful use of language? Are expressions conditioned by the nature of things to which they refer, outside ourselves, or can they reflect only our understanding of those things? Do expressions represent states of being within ourselves? Do we use expressions to modify states of being within each other? Is it a criterion of success that expressions should feel right in conjunction with our further perceptions of our world?

The following notes set out a framework for thinking about these questions. Two main concerns will influence the further discussion. First, there is the ambition that things other than humans, such as computers, should be given the ability to use human language. Secondly, if this ambition is attainable, then computers ought to be able to use language expressions in the form of pictures as well as words.

PARTS OF LANGUAGE

A broad understanding of language as a means of communication between intelligent beings, people, which embraces all forms of expression as well as interpretations within human minds, can be considered in two main parts:

(*a*) *expressions* that are realised outside people and are passed from one person to another; and

(*b*) *interpretations* within people, that are the semantics or meanings of expressions as felt within ourselves.

Expressions

An expression has:

a *realisation* in some substance — as in the medium of voice (sound waves), text, or drawings; which has

a *structure* that governs composition and decomposition of realisations; and

a *function* to differentiate the realisation from other realisations; and it might have

mappings to other different realisations, possibly in other media — different realisations that evoke similar interpretations.

Interpretations

An interpretation is:

an *abstraction* from an expression, that is conditioned by an expressive environment and is determined by a state of mind — a concept of a person or people; that refers to

an *individual* as something within (and conditioned by experience of) the interpreter — constituting knowledge of 'the world of people'; and evokes

logic as the reasoning ability that is within (and is part of) the interpreter, that relates one individual to another, and determines satisfactory correspondence with expressions; which might be realised as

mechanisms that serve as externalised representations of logic, outside the interpreter — as in a mechanical or electronic device outside a person.

Individuals

Of all these parts of language, *individuals* might be in most need of further explanation. These are conceived as states of things within people which are formed out of the conjunction of whatever constitutes the mind and people's sensory experience (through sight, sound, touch, taste, and smell) of an external world. Individuals can be thought of as people's notions and as primitives by which we can distinguish things that occur in our experience. Logic is used to develop structured arrangements of individuals, to form human representations of the world. Unique arrangements then constitute *individuality* which, through the behaviour which it causes people to exhibit, distinguishes one person from another. Beyond this conceptual schema, it is not necessary to know of individuals as actual and particular things;[3] it is enough to think of them as conditioning a person's behaviour and, thereby, conditioning human knowledge. Individuals can be private to persons. If individuals could be externalised, there would be no need for language. The role of language, manifest through expressions, is to allow people to reach into and interact with each other's individuals. Uncertainty about correspondence between different people's individuals explains why we go on talking.

This meaning for individuals differs from the orthodox technical usage of this term, which refers to external objects differentiated by their properties (or relations). However, my meaning relates to common usage. It does so in the sense that it accepts individuals as indivisible and unanalysable things, and it accepts individuality as indicating differences between things. In this sense, we can consider individual people and, by implication, individual objects.

All the parts of language that have been outlined here are necessary to the existence of language. For language to be effective, using any formal environment of expressions, abstractions from expressions and the use of logic that determines arrangements of expressions have to be public. The success of language then depends on particular interrelationships between public and private parts of language. This interdependence explains why language has to evolve as people evolve.

DEFINITION AND CONSEQUENCES

This framework provides a rough definition of language which is intended to accommodate all modes of expression, extending beyond verbal or written language. My intention is to show that the use of structure in expressions and logic in interpretations, which is commonly the focus of linguistic studies, also applies to drawn expressions. Structure and logic applied to pictures may differ but their role in making pictures serve as linguistic expressions is similar.

One striking difference, however, is that pictorial expressions require us to look more closely at the structures of their realisations; at pictures as externalised and tangible objects. Linguists usually focus their attention more directly on logical structures of abstractions. They can do so because they can assume that text realisations, depictions of words, exhibit differences that correspond to differences they employ in their abstractions. In the case of pictures, we do not yet have a regularised and widely acknowledged correspondence between graphical realisations and logical structures. When linguists wish to attach significance to variations in the ways in which words may be depicted, the graphical formation of words, then they too have to focus more attention on the realisations of expressions.

As with any definition of language, my definition poses deep problems and I will now acknowledge some of these problems.

Abstractions and individuals

Individuals are our own thoughts formed in mind, conditioned by our experience of the world. They contribute to the individuality of each one of us. Abstractions are things that we take into mind from expressions and they condition our use of expressions. We use expressions to represent individuals that constitute a discourse domain, to convey knowledge between people.

In any discourse, we cannot know that different people use precisely the same abstractions and refer to the same individuals. We cannot know exactly what each other knows of our world. Moreover, we also cannot know whether these abstractions and individuals are different. The underlying difficulty here comes from my acceptance that what we know is conditioned by the kind of being that we are. We cannot stand outside of ourselves to discover what we are, at least not those parts of us that condition our abstractions and individuals.[4] However, it seems plausible to assume that as similar beings, with similar senses by which we experience our world, we do have similar abstractions and individuals.

This similarity can be explained by a more concrete example. Our senses do allow us to observe certain parts of ourselves as though we were standing outside those parts, and we can observe that we have hands. We can observe that we all have hands (barring accidents). We can then think of all hands as being the same as each other and different from other parts of ourselves, for purposes of knowing when we are looking at a hand. However, we can also think of my hand as being different from any other hand, for purposes of knowing whether this particular hand could have been involved in some action. For example, my hand or fingerprint might be identified in a burglary. Things like hands, which can be both the same and different, can be thought of as being similar. It is this sense of similarity that I am applying to abstractions and individuals. My use of my abstraction from the word 'hand' will in some ways be unique to myself and in other ways it will be the same as uses of corresponding abstractions within another person or in many other people.

Reference

Individuals are commonly thought to correspond to real things in the world which we experience. I can use my logic to compose individuals into my mental picture of some situation, and I might use my abstractions from an expressive environment in order to realise my mental picture in the form of a drawing. If I mean to depict my experience of some real situation in the world outside myself, then the common expectation is that my depiction of individuals should correspond to the real situation as it actually exists, independently of our thinking about it existing.

The problem is how can we know of such objective existence? We commonly behave as though we want to have independent reference points in reality by which we can verify our various individuals. Put more strongly, many of us would like individuals to be such independent beings. We are not able to establish such a position for a simple reason, as noted above. To establish correspondence between, or equivalence of, an individual and an objective thing, we would have to stand outside ourselves and look at our individual and the thing and determine that they were objectively the same. Even if we try to be very subtle about this ambition, ultimately it is absurd.[5]

This ambition also is unnecessary. Consider the possibility of reality in the form of a world that might be composed of parts — objects, actions, situations — and we are some of those parts.[6] Such a reality can be accepted without our having to know what it actually is. I do not need to know what any part of the world actually is, outside of my experience of it. My senses enable me to experience reality. I assimilate that experience as arrangements of individuals which become part of me, contributing to my individuality. As an intelligent being, I can exercise my logic on my individuals in order to compose my pictures of the world as it might have been, as it might be now, and as I think it will be.

This subjective view of reality is likely to produce idiosyncratic worlds. It is necessary to reconcile my individuality with the subjective views of other people; we have to cohabit in the same reality in order to survive as a community of people. Here is where language plays a crucial role. It enables us to reach into each other's subjective views of the world and to reconcile our differences in an inter-subjective view of reality.[7] We have to behave as though we are using the same abstractions and individuals, and we strive to do so by adjusting ourselves to each other's expressions and responses. By doing so, I can relate myself to other people's experience and extend my own view of reality to embrace human knowledge about such things as science, art, and religion.[8]

Thus, our individuals can correspond to a reality, but this has to be a humanly constructed reality. This position simply ignores the question of what a separately existing objective reality can be, but it accepts the possibility of that reality influencing our inter-subjective reality. Our experience has to be built on something that exists 'out

Language, Logic, and Intuition

there'. The point of this position is that we cannot know what objective reality really is, and we cannot call on objective reality to verify our inter-subjective reality. Instead, our reality has a fluid existence. It depends on contributions from each of us, which, in turn, depend on the kind of being that we are. Our reality has to be maintained by us and it evolves with us.

Logic

I have already said that logic can be considered as reasoning mechanisms for constructing representations of human understanding. These mechanisms are part of us and, by using them, they make us exhibit intelligent behaviour. By equating logic with mechanisms, I do not mean to imply that we can describe the whole of human logic in terms of mechanisms that can be realised outside people. However, I recognise that people are at present embarked on a project to realise human logic in computers, and some see no limits to that possibility. In that context it is useful to think both of people and computers as being machines, to consider the kind of mechanisms that might explain human behaviour, and to consider the feasibility of realising those mechanisms in computers.

Logic, in my view, is not embraced only by formal realisations of logic, as in classical logic, variants of non-classical logic and, indeed, mathematics. These all are attempts to externalise and regularise some aspects of human logic. If logic is considered as mechanisms for establishing truth conditions, then it is evident that people have mechanisms for arriving at truths which cannot be accommodated by any current formal logic. For example, we do not have any such logic for the spatial modelling of deep aesthetic sensitivities, as expressed in the architecture of buildings.

The problem here lies with the way in which we establish truths. We generally think of something as being 'true' if we know that it exists or we know of other things that can be made to correspond to it. If this condition is not satisfied, various formal logics treat such an outcome as failure (it does not exist), or not proven (it might exist), or not yet proven (we might expect, in time, to discover that it does exist). Formal logics have to be manifest in the use of expressions that represent whatever might be known in mind (see later). This connection between logic and expressions focuses attention on language.

The problem is that we do not have a formal logic that can embrace all that can appear in people's natural language expressions. This is evident in the case of pictorial expressions and in symbolic expressions in the form of words. The general condition for truth holds only if we accept that there are things we have in mind which we are able to relate to things that occur in expressions, and that those things in mind do not appear in expressions. People have reasoning mechanisms which we cannot explain outside of people. These human mechanisms are used perhaps unconsciously within people, as part of people, and they contribute to the establishment of truths that are accepted between people. Truths arrived at in this way are people-centred truths.

The point is that we can have no solid basis for differentiating between truths which are people centred and truths which are arrived at by means of formal logic. Any formal logic is a realisation of people's understanding of some aspect of human logic. Ultimately, the truths of aesthetics and of physics are the same.[9]

This position does not deny the importance of formal logic. It does, however, pose awkward questions for any separation between formal and human logic, especially if this is intended to lead to separately functioning beings — as in the separation between people and computers. People who employ formal logic are able to moderate their results by also employing their own human logic to arrive at humanly acceptable truths. It is not important to maintain a separation and people, even formally trained people, do use formal logic quite informally. Computers, on the other hand, as reasoning mechanisms realised by people, outside people, can perform only those operations defined exclusively on formal logic. As representations of human reasoning mechanisms, computers are not the same as human logic, and they represent certain people's understanding at some point in time. Operating independently and detached from human logic, we should expect computers to offer us a very limited set of truths and even fewer humanly acceptable truths.

Expressions
Expressions are the tangible realisations of language. These are humanly devised artefacts produced with expressive intent. They are intended to evoke abstractions in mind which refer to individuals and are amenable to logic. These abstractions and individuals are the interpretations of expressions.

What distinguishes expressions from any other artefact? On a broad definition, very little. All artefacts can be considered to have some expressive intent. Even mechanical devices can be regarded as referring to individuals, representing knowledge about physical laws in different people. When these devices are styled for a market, as in the case of motor cars, then they exhibit evocative intent more obviously.

Language, Logic, and Intuition

However, a definition is required which is more closely allied to my earlier definition of language. It needs to be allied to the use of logic aimed at formulating statements about what can be. For example, we want to identify the kind of expressions that can represent what we think a car can be before we have a car, and what a car can imply in relation to other things. These are not solid criteria but they are suggestive. This leads us to expressions in the form of words (either spoken as sound waves in time, or written in one-dimensional space), pictures (drawn or painted in two-dimensional space) or physical models (sculpted or assembled in three-dimensional space). This discussion will focus on written words and drawn pictures.

With this focus, it now becomes necessary to identify what distinguishes written words as a form of expression, and what distinguishes pictures. In both cases, we are dealing with realisations in the form of 'marks on paper'. We have to consider the properties of those marks and their relations in compositions of marks, which then receive interpretations when they are used in language.

Written words and word compositions can be considered as linear strings of characters, punctuated by blank character spaces, commas, etc. Words are bounded by blank spaces and are differentiated by their unique combinations of differentiated characters. Coupled with an ability to make and manipulate objects that exhibit these properties, we then have an expressive environment for realising written words — a writing production environment.

We have abstractions for particular words and word sequences which signify objects (nouns), actions (verbs), and conditions (which nouns are subject to which verbs and in conjunction with which other nouns and verbs?) which define syntactic situations composed of words. We make these situations refer to individuals we have in mind so that we can relate a written text to a discourse domain. When we do all this, then written expressions serve as realisations of language.

Pictures in the form of line drawings are composed of lines. Lines are differentiated by style, curvature, angle, and intersections with other lines. Length is conditioned by the scale attributed to a two-dimensional drawing space. Chains of lines can be defined by attachment states at intersections, and can be differentiated by unique combinations of values for their lines. Again, coupled with an ability to make and manipulate objects exhibiting these properties, we then have an expressive environment for realising drawn pictures — a drawing production environment.

We have abstractions for particular arrangements of lines and chains, and for operations on lines which signify objects (shapes), actions (translate, rotate, scale)[10] and spatial relations (connectivity, proximity, similarity) which define syntactic situations composed of lines. We make these situations refer to individuals in mind so that we can relate a drawn picture to a discourse domain. Again, when we do all this, then drawn expressions serve as realisations of language.

Definitions like these are rough and unstable, but they can help to explain our traditional systems for producing expressions. In these traditional systems, people decide what to produce and then use their knowledge of an expressive environment to execute a production. People, as intelligent beings, are able to mix expressive environments so that we are able to combine the expressive convenience of text and drawings. We do so by deciding on mappings across these expressive forms, which depend on our interpretations.

The point here is that definitions like these (implicitly or explicitly) underlie our traditional systems and they provide the basis for new systems. New systems are promised in the form of computers and the promises include the possibility of externalised machines that can understand our expressions. In the next section of this chapter I will explain why I think that possibility is not likely to be fulfilled. Instead, we can expect new systems for producing and operating on expressions, as indicated in Figure 2.2 and explained further in Chapter 7. These systems will offer new expressive environments and they will operate with received mappings to expressions in our more familiar expressive environments. Irrespective of whether these systems possess independent intelligence outside ourselves they will have an influence on our future intelligence.

Deeper into Logic and Expressions

Logic has been considered as reasoning mechanisms for relating individuals that people have in mind. Expressions, through abstractions, refer to individuals. To illustrate the relationship between logic and expressions, from a human viewpoint, we will now look at a sample series of statements. Individuals in the mind of the utterer are denoted without quotes, and expressions from the utterer or a respondent are denoted with quotes.

(i) P $\rightarrow$ 'Q'
(ii) P $\nleftarrow$ 'Q'
(iii) P $\rightarrow$ 'Q', 'R'
(iv) P $\rightarrow$ 'Q', 'R', S

Language, Logic, and Intuition

The first statement (*i*) consists of an individual, P, which produces the expression 'Q' — this means that the expression serves as a representation of P, where P stands for individuals that will be similar but different in the minds of the utterer and respondent. From previous discussion, we cannot know what any particular individual is, so we cannot know whether the expression represents all that might be known as the individual (*ii*) — the reverse relationship does not hold. The individual P might also be known to the utterer as a composition of individuals that might be indicated by the expression 'Q' and 'R' (*iii*), without contradicting the first statement. It is also possible that this composite statement requires a further individual, S (*iv*), which is not expressed in the same notational form as 'Q' and 'R', to qualify the conjunction of 'Q' and 'R' and make this expression humanly acceptable as a representation of P.

Invisible individuals
The most interesting point to note in this illustration is that the individuals, P and S, would not appear in the expressions realised between the parties to this discourse. The expressions would include only 'Q' and 'R'. Yet the individuals plus mechanisms for realising their logical implications are necessary to any human interpretation of the expressions. The people who are the parties to the discourse need to be thought of as parts of the statements — the expressions without the people would be meaningless. This association of people with expressions points to the existence of human logic operating within people. Given that people cannot see each other's individuals, we can characterise human logic as including an ability to respond to what is not known, at least not in an externalised and formal sense. This ability is evident in people's uses of expressions and in normal human behaviour.

As an elaboration of this point, consider the use of S in the last statement. If the individual P is what is known to the utterer as a house, then 'Q' and 'R' might refer to parts of a house, such as walls and roof, as depicted in Figure 2.3. If, say, 'R' could also denote other things then S might be what is known about the necessary conjunction between 'R' and 'Q' to make 'R' a roof. This conjunction might be known in a manner that cannot be expressed (perhaps not in a prior form that can be related to the particular context presented by the current statement), yet it will have to be known to the person doing the interpreting. Furthermore, the utterer of the expression has to rely on the respondent also possessing that knowledge — if the respondent shows that this is not

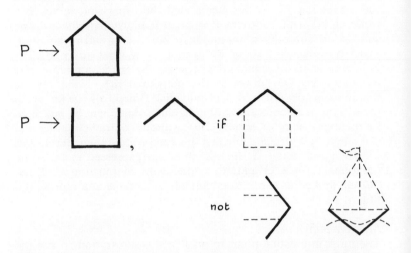

Figure 2.3 *Silent Conditions*. P might be what is known as a 'house', which might be known as being composed of the parts 'walls' and 'roof'. The roof is a roof if it sits on the walls, and not if it is used as an arrow or a boat; such conditions on the expression are known to the utterer, but are not revealed in the expression.

the case, the utterer has to say some more. Here we have an example of a lot of knowledge being required about something that is not realised within a composite expression. Such a situation tends to be evident in the case of expressions which are in the form of pictures, but it is also normal in text.

As a further general point, the quoted denotations in the illustration can be realised in words, or pictures, or any other form of expression without altering the logical relationships. Note also that P and S might be thought of as real world objects, as though individuals might be known as having some independent existence, without changing the logic. The main reason for not doing so is to avoid the implication of objective authority for the expressions.

Formal logic
When we restrict ourselves to formal logic we have to accept the condition that all operations have to be defined only on expressions outside people. Thus, the only legitimate terms in the illustration are those that are shown with quotes, 'Q' and 'R'. To restore logical functionality we have to replace the individuals with the expressions 'P' and 'S'. When we do so we are in effect expanding the statements

whilst remaining silent about the further individuals needed to make the statements humanly acceptable. Someone has to declare the logical implications and those implications will be conditioned by that person's individuals represented by 'P' and 'S' as well as 'Q' and 'R'. Other parties to this discourse also have to recognise these expressions as referring to their own similar but different individuals. Once all this is done, then an external logical machine might be made to operate on these expressions, using 'true if' or 'if true then' inferencing mechanisms to establish that 'P' refers to a house with a roof.

What does this revised illustration tell us about formal logic? Does it tell us something about any attribution of intelligence to logical machines realised outside people, to computers? The most obvious observation is that computer operations are dependent solely on received expressions and they apply only to expressions.[11] The formal logic that defines a computer has to exist only in expressions in the form of words and (conceivably) pictures, or more formal notations or bit patterns. Since we cannot replicate individuals in expressions, we cannot give computers our own individuals.[12]. It is just conceivable that a computer may possess its own individuals, conditioned by the kind of thing that computers are, but those individuals are likely to be unfamiliar and inaccessible to people. It then follows that a computer cannot use human language to communicate with people.[13] Even if it utters expressions that are familiar to us, at best its abstractions for those expressions will refer to radically different individuals or, at worst, it will have no interpretations at all. This absence of language has the implication that computers have no means to convince us that they are intelligent. Nor, conversely, can we convince computers that we are intelligent.

Dumb machines

At this point it seems appropriate to drop the idea that computers can be intelligent. Of course, any computer does exhibit the ability to produce expressions and it is able to do things that have the effect of changing expressions. It can receive sequences of expressions from us which can have the effect of controlling its exercise of its abilities. All this does not amount to language. Instead, it amounts to a machine for producing and altering expressions (Figure 2.2). These expressions then play a part in the language that passes between people when we use the machine. People give interpretations to these expressions and we can use the machine to produce expressions that reach into our various interpretations. Thus, we come to a characterisation of this position as people's intelligent use of dumb machines.

KNOWLEDGE

To round off this discussion, I will now address a number of concepts that are usually associated with computing, and set them in the context of my understanding of language. I will consider knowledge, representations, actions, and responsibility, and I will suggest how these concepts apply to human activities, as in art and science. This covers a very large spectrum of concerns which can be dealt with only briefly here, in order to point to some interesting consequences of how we think about things.

Computers are strongly identified with knowledge — we have knowledge engineering and knowledge representations, processors, and bases. In all these cases, the assumption is that knowledge can be something externalised from persons, with an objective existence, and that we can know what instances of such knowledge are. I prefer to think of instances of overt knowledge as representations of knowledge within persons, with the implication that the representations are not the same as human knowledge. In my discussion of language I have said that people know things and knowing refers to our individuality. Knowledge can then be thought of as logically structured arrangements of individuals, employing logic within humans and conditioned by the kind of being that humans are. I have also said that we cannot know what individuals actually are, which implies that we cannot know what knowledge is, and we cannot know that some particular knowledge is absolutely correct. This general position is rehearsed in Figures 2.4 and 2.5. How can this position be reconciled with the idea that knowledge can usefully reside in computers?

Immediately, my understanding presents a paradox. How can I say that we cannot know what knowledge is, if not knowing refers to some state of individuality that exists in response to something that we call knowledge? Our individuality constitutes knowledge and, if this position feels right, we can know that fact about knowledge. By the same argument, we can say that thinking about anything is knowing and, as thinking involves our individuality, knowing is feeling — knowledge feels good and right, or bad and wrong. This position appears to explain much of human behaviour, as in poetry and in more mundane activities. It does not, however, tell us how knowledge can exist in a manner that is detached from people, as in computers.

Language, Logic, and Intuition

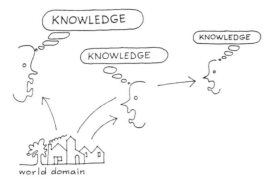

Figure 2.4 *Intuitive knowledge*. A person's knowledge of the world can be thought of as making the person intelligent, but if we had no means of exhibiting knowledge we would not know. Knowledge wholly contained within persons would leave us with the kind of intelligence we might associate with dogs and cats.

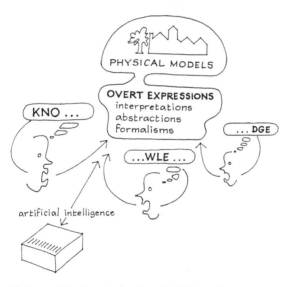

Figure 2.5 *Externalising knowledge*. People's ability to give concrete expression to at least some of our knowledge is essential to our appreciation of human knowledge and the development of that knowledge. This ability rests on the way that we are able to link interpretations to abstractions from formalisms used in expressions. Expressions are partial representations (or models) of knowledge and we cannot know whether we use precisely the same interpretation, which explains how people can legitimately read different meanings into the same expressions. We then face the question: how can a machine be used to generate and interpret human expressions?

Kinds of knowledge

Consideration of the conditions that might be necessary for detached or objective knowledge prompts us to try and differentiate between kinds of knowledge. I will outline three broad kinds, all probably being aspects of the same human knowledge. It is only the last of these kinds which offers the possibility of recreating knowledge in a detached form in computers.

First, we have an innate sense of knowing and intuition. This refers to our individuality and logical abilities within ourselves, which contribute to our existence as living beings. Our innate sense tells us how to interpret our experience and lets us feel when things are right or wrong. Intuition is our knowledge of ourselves in our world in a form we cannot externalise objectively in overt expressions. Intuition is knowledge by virtue of the fact that we use it to condition our actions in response to experience.

Secondly, we have common sense. This refers to knowledge shared between people which constitutes our inter-subjective reality and is evident in our shared practical ability to do things. Common sense requires people to be able to externalise what they know, but this knowledge is not fully overt in that people do not necessarily know how their externalisations work. Typically, common sense is expressed informally in natural language, and it employs expressions that are interpreted more directly through our animal senses (including sight, sound, touch, taste, and smell). This use of animal senses may be necessary to link a person's innate sense of knowing with shared common sense.

Thirdly, we have formal knowledge. This refers to knowledge about formal structures of abstract worlds which is used to construct overt representations of people's experience. These representations appear as expressions of some formal language, or formalism, so that expressions can be thought to correspond to individuals and logical operations on arrangements of individuals within persons. Typically, formalisms employ classical logic and its derivatives which, as I noted earlier, are formally treatable aspects of human logic. It is this kind of knowledge which is used mostly by scientists and theoreticians when they seek to explain common sense and innate knowledge. This is also the only kind of knowledge we are able to recreate in computers.

Having drawn these distinctions, I should restate that these kinds of knowledge are probably integral parts of the same human knowledge. Clear demarcations are elusive. For example, formalisms cannot be complete in the sense that externalisations do not reveal explanations for the functions they represent,[14] nor do they

Language, Logic, and Intuition 43

include explanations for how they are known and used within a person.[15] If we do not know something in our common sense or intuition, then using a formalism will not get us to a state of knowing it. This condition applies to the work of scientists just as it does to other people.

Representations and actions
Overt representations of knowledge are evident as externalised expressions. However, a representation of something is not the same as the thing it represents, otherwise it would be the thing. A representation is subject to the filtering effect of interpretations from something as it can be known in one environment, as in a person, to something as it can be known in another environment, as in a computer. Interpretations are conditioned by what each environment can do.

This observation ties in with my earlier discussion of logic and expressions, formal logic, and dumb machines. Using computers, we have to restrict ourselves to using formal logic but, given the distinction between representations and things that are represented, we can choose to represent anything in a computer. When we do so, we will have no objective criteria for establishing the 'correctness' of a representation with respect to something outside the representation but, provided we are satisfied that we know what the computer can do and that its functionality corresponds with what we want to do, we can decide to accept the representation. This is a heavy but inescapable proviso.

Computers are considered to be general purpose machines. This means that they offer a representation environment in which useful logical representations can (potentially) be constructed for anything, and they can be made to exercise their functionality meaningfully on any representation. This claim of generality rests on the possibility of separating representations, as passive reflections of other things known outside a system, from actions, as active intentions of people for those things.

Representations, as passive reflections, have to be defined in terms of things, and only those things, that are known to constitute the representation environment — the system's abstractions (or entities) and the system's logic (formal logic), in terms of language. To be general, the system must be able to receive expressions from any person, and map these to its abstractions to produce its representations of things that the expressions describe. Ideally, these representations ought to reflect all possible readings of expressions from different people, so that many expressions can result in interrelated representations within a single representational environment. Such representations should then be available for any actions people might wish to perform. This ambition entails ambiguities and contradictions which do not matter until we ask the system to do something with a representation.

Actions, as active intentions, refer to things that people want to do to other things. If we want actions to be executed by a system then the actions have to be described to the system. An action has to be defined on the functionalities that are available within the system, such as go (from here to there) and replace (this with that). Moreover, a condition of any system that relies exclusively on formal logic is that it can resolve actions only along linear paths through a representation.[16] Ambiguities and contradictions have to be filtered out. Thus, the definition of an action also has to include references to partial representations (or views) of the thing on which the action is to be executed, ensuring that those representations convey only single values as and when required by the action. We can think of actions as exercising responsibility for identifying those things on which they act.

To preserve generality, a system must be able to receive descriptions of actions from any person. The expressive environment that links a person with the system must include expressions that invoke functionalities of the system, plus a syntax for associating those expressions with further expressions that refer to representations. Ideally, such an expressive environment ought to feel like natural language. For reasons that I have outlined before, this ideal is probably not attainable. We are faced, instead, with a strictly conventionalised programming language for controlling mechanistic operations within a computer. Generality then depends on people's ability to perform translations between natural language and a programming language. Irrespective of future advances in high-level or natural-like programming languages, this need for translation will persist. Alternatively, but improbably, people might come to adopt formal logic as an exclusive determinant of natural language.

Responsibility
Responsible actions are matters for judgement by people at whom the actions are targeted, and those people then attribute responsibility to the authors of the actions. This is a fairly orthodox people-centred and democratic position. Complications can arise when people seek objective criteria to validate their judgement but, typically, conclusions are moderated by opinions of human experts which are, once more, people centred.

A more serious problem arises when the authors of actions use devices that contribute to those actions, when they use computers. Who, or what, bears responsibility for those actions? From all that I have said so far, it should be clear that users of computers have to bear responsibility for their use of computers. They have to do so even if they regard their computers as possessing knowledge which tells them what action to present to other people.

The argument I used to explain why computers cannot share human language also supports the conclusion that computers cannot share human knowledge. At best, computers can perform certain operations on certain representations of human knowledge, and they can do so very effectively, but this cannot be the same as people exercising their own knowledge. Computers, therefore, cannot share human responsibility for actions.

This conclusion persists even when people use computers which have been programmed by other people. Indeed, that is the normal situation for nearly all users of computers. Having been programmed does not alter the fact that the behaviour of computers is determined exclusively by formal logic — and we would be even less confident if we could make them work by any other logic. We are back, once more, at the fundamental point that computers and people are different kinds of things. Consulting a programmed computer is not the same as consulting the person who programmed the computer.

This position on responsibility explains why the most widely successful applications of computers deal with tasks that are readily understood by people, such as word-processing, arithmetic in spreadsheets, and production of drawings. Responsibility for what words are processed, what quantities are calculated, and what drawings depict rests with users in a manner easily controlled by users. Progress towards new and more ambitious applications is possible and may be desirable. However, such advances may turn out to be more dependent upon spontaneous evolutionary changes within people, and not only upon ingenious technological developments.

Open and Closed Worlds

To conclude this chapter, I will now illustrate its substance by sketching two contrasting views of knowledge. The first focuses on computational symbolic logic which employs the closed-world assumption of classical logic, and is the logic we are able to implement in computers. The second is a sketch of human intelligence operating in an open world.

Computational logic

When we expect computers to do anything useful, like resolve a linked succession of expressions, we are faced with the closed-world assumption. This assumption says that all propositions have to be resolvable only as true or false, with reference only to what is already known. This position can be characterised as follows:

> If you ask me about something
> and I don't already know about the thing
> then I must reply that it does not exist.

The closed-world assumption does not accommodate:

> perhaps, possibly.

From a common sense point of view this position appears absurd, and it probably is. The position becomes a little less absurd if the reply in the above statement is read as meaning 'does not exist, as far as he knows'. A computer program can be made to mask its closed-world assumption when it finds a truth value that is false by getting it to say:

> I don't know.

This leads to intractable computational problems. If a program is being asked to perform a task that consists of a number of sub-tasks, and if it is expected to use the result of one sub-task as part of another sub-task, and if it has to do so successively through several tasks, then it will be in trouble when it reaches a 'don't know' conclusion anywhere within the succession of tasks. The computer could ask for help but, for complex programs, the possible escalation of situations requiring some kind of user interaction becomes unmanageable.

One might expect a sophisticated computer program to say (to itself or to a user):

> I don't know yet,
> but I'll carry on with what I have got meanwhile
> and I might find that I do know later.

This would be equivalent to the 'perhaps, possibly' position and would take us outside the closed-world assumption. It illustrates a strategy that is well established in common sense, but among formal theorists it remains a serious topic for speculative research. Formalisms do exist that are based on a variety of non-classical logics, including intuitionistic and probabilistic logics,[17] but so far they do not form the basis of useful computer programs, and we do not know whether they ever will.

Human intelligence

As a contrast to closed worlds, the following sketch illustrates

Language, Logic, and Intuition

human intelligence aimed at geometry depicted in drawings. We will consider Euclidean geometry which describes shape properties, employing line entities and their relationships. Geometric expressions are made up of compositions of lines, and these might then be interpreted as properties of other objects outside the domain of geometry. We can represent our knowledge of Euclidean geometry in the form of inference rules and a fragment might appear as:

We have a triangle
if:
we have a part of a two-dimensional plane
and:
it is wholly bounded by straight lines
and:
the quantity of lines is three.

Note that the first part of the expression is a proposition and the succeeding parts are conditions that need to be fulfilled in order to substantiate the proposition. Strictly, we should read all these conditions as further propositions which should each be followed by their own further conditions, and so on, until we have defined all the parts of our definition of a triangle.

Assuming that we can represent this knowledge in an inference machine and assuming that the machine can recognise the presence of lines in some drawing space, we can then expect the machine to respond to a variety of questions about the presence of triangles in drawings, such as:

Does the drawing contain a triangle?
How many triangles?
Is this part of the drawing a triangle?
If I remove this bit, is the rest still a triangle?
If not, why not?
What is a triangle?

It is feasible to make computers behave in this way and this looks like intelligent behaviour. However, there remain important limitations to this manifestation of intelligence. As explained previously for closed worlds, it is possible to ask only those questions that have answers which are already explicitly and wholly contained in the rules. The answers can refer only to the domain that is described by the rules; in this example, to geometric properties of objects that occur in drawings and not to any other properties of further objects that drawings might depict.

People's intelligent behaviour appears to extend beyond these limitations. People appear to know what they do not know:

> Suppose that my mind was empty of knowledge about Euclidean geometry and triangles, and suppose that someone else started to talk to me about triangles occurring in a drawing. I might acknowledge that the other person had some notion of something which he was calling a triangle. Without revealing my ignorance, I would nod my acknowledgement and might even voice responses that he would accept as sensible, encouraging him to go on speaking. As the conversation proceeds, I could piece together clues that would gradually explain to me what he meant by triangles, without my asking any direct questions. My acquired understanding of triangles might then survive further conversations with yet more people who know about triangles. Only later, after I had already come to know about triangles, might I come across a formal definition of triangles.

This ability to acknowledge absent knowledge and continue functioning without previously having had to *anticipate* the absence is part of the common experience of people and we know no way of representing this ability in computers. To continue with the example:

> During the course of my acquiring knowledge about triangles in the context of drawings, I would also acquire knowledge about the role of drawings in depicting other things, the representational properties of drawings. Consequently, I would link my understanding of triangles to my understanding of drawings so that I could interpret geometric properties of triangles into properties of other things depicted by drawings. This interpretive ability would depend on my experience of those other things, my acquisition of knowledge through my various means of sensing those other things.

Again, acquisition of interpretive ability without previously having had to *anticipate* what things might need to be interpreted is part of the common experience of people and we know no way of representing this ability in computers.

It is possible to argue against this presentation of human abilities and the limitations on corresponding abilities in computers. Counter arguments can be expected from artificial intelligence and related fields, and the debate quite properly becomes deeply philosophical.[18] It would be presumptuous to predict a conclusive outcome; perhaps we do not want one. However, the position I have presented here is likely to condition all practical computer systems in the foreseeable future.

SUMMARY

This chapter has outlined my philosophical position on intelligence, language, and logic, with consequences for the way we think about knowledge and our ambition to replicate knowledge in computers. This position is relevant to my further discussion of experience of CAD in the following chapters.

Human intelligence, as exhibited in our use of language, has been described as part of being human. Our view of intelligence has to be biased in favour of people, and more so among like-minded people. Our experience of all things conditions our intelligence; we recognise our interdependence with other things, and we use our intelligence to help us survive. We can conceive the possibility of other non-human intelligence but, outside behaviour that we can recognise as matching our own intelligence, we cannot know what any such other intelligence is.

Language has been described in two parts: *interpretations* people have in mind, which shape the individuality of persons; and *expressions* realised in some externalised form, which pass between persons. Interpretations constitute knowledge and rest on states of mind within persons, conditioned by human experience and the operations of reasoning mechanisms in mind. This understanding of interpretations does not require us to know what the mind actually is or how it works — we can simply accept that it is part of being human. Expressions are realised from the mind and become things in the world that is experienced by people, as representations of knowledge. People can then read expressions by taking abstractions from these things into mind, and reconstitute them as knowledge. Interpretations are presumed to be similar in similar beings, leading to our inter-subjective reality, and expressions are the means by which we reach into each other's minds.

Reasoning mechanisms in mind can be considered as being conditioned by human logic, used to establish our inter-subjective reality. This reality applies in a uniform way to our knowledge about our world, including both science and art. Formal logic can be viewed as attempts to externalise aspects of our inner logic, and so we have classical logic, variants of non-classical logic, and mathematics. Any formal logic, when it is used in conjunction with human interpretations, remains inseparable from people. This condition holds irrespective of the form of expressions used to represent knowledge.

Computers are artefacts that are made to be externalised reasoning machines. However, they operate on received mappings between expressions, without human interpretations, and they cannot share our language. It is, therefore, misleading for us to think of computers as being intelligent. Instead, computers can be thought of as devices for producing and manipulating expressions, when people exercise intelligence in making them do these things. Computers can be made to offer new expressive environments by which we can reach into each other's knowledge, and we are interested in using them to improve our ability to express design knowledge. Computers have now become part of our reality, and increasingly they will have an effect on our intelligence.

Notes to Chapter 2
1. The position set out in this chapter is likely to attract severe criticism, especially from some members of the computer science and AI community. It includes assumptions which are not likely to be acceptable to those who have ambitions for practical AI applications. My purpose is to consider the conjunction of computers and design from a human rather than an AI perspective. I wish to discover what we might want computers to do, rather than circumscribe our understanding of human behaviour in terms of what we already know we can make computers do.
2. Recent work with computational linguists refers to collaboration on ESPRIT Project 393: ACORD, which focuses on natural language text and graphics interrogation of knowledge bases (Klein 1987) — my views presented in this chapter are not necessarily shared by other collaborators.
3. It is tempting to see a parallel between my understanding of individuals and what Dawkins (1978) calls 'memes', which he describes as a sort of mental counterpart to genes that determine the behaviour of their host 'gene machines'. Individuality, like meme structures, can be thought of as being subject to rearrangement in response to experience and we recognise this as intelligence.
    My reference to individuals might also bear some similarity to Maturana's concept of mental structures that produce physiological behaviour which is manifest through language (Winograd, 1980; and Winograd and Flores, 1986). He argues for the possibility of convergence of states of mind, without requiring correct representations (or models) of the world.
4. This inability to stand outside oneself can be regarded as a generalisation on Godel's theorem (Flew, 1984), which says that a system cannot look at itself and which sees persons as systems.
5. This condition on objectivity is fairly widely recognised among philosophers such as Putnam (1978) and Johnson-Laird (1983), and it is evident, for example, in both Wittgenstein's earlier and later works (Pears, 1985).
6. Our being within and part of the world that we experience accords with the position postulated by Heidegger (Steiner, 1978; and Winograd and Flores, 1986).

Language, Logic, and Intuition 51

7. If lesser animals do not need to use language to support intersubjective reality, the explanation might be that their in-built logic is not sufficiently powerful to enable them to build destructively divergent subjective views of their world — they are not intelligent enough to be mad.
8. The need to develop an account of language that can cover a wide spectrum of knowledge, embracing science, art, and religion, is what appears to have motivated Wittgenstein's later work (Pears, 1985).
9. The idea that even physics is founded on people-centred knowledge is fairly widely shared by modern physicists. It is supported by Capra (1983) and, perhaps more authoritatively, by Einstein (in Johnson-Laird, 1983) when he says: 'Physical concepts are free creations of the mind, and are not, however they may seem, uniquely determined by the external world.'
10. There is an interesting divergence indicated here between words and drawings. Actions in the former case occur in interpretations and they condition edits in the production environment. Actions in the latter case are more apparent as edits in the production environment, and they condition interpretations. This might indicate that the former case needs to be viewed as a product of the latter; text, as depiction of words, is a special case of drawings.
11. A computer's dependence on received expressions is a condition which holds even when computers are used to control mechanical actions.
12. It might be argued that we can build certain human sensitivities into computers so that they might develop individuals similar to our own. We might, for example, give a computer a sense of touch, and we might tell it that a certain touch experience is 'cold' and another is 'hot', but how would it get to know that being on fire is not a good idea?
13. If we were to consider utterances of computers as evidence of language, we would have to consider where such utterances come from, how they are motivated, and with what intent. At best we might conclude that these utterances come from other people who use computers to realise their own expressions.
14. As an example of incompleteness of formalisms, consider the difficulty that Russell and Whitehead (1910) experienced in trying to uncover how numbers work.
15. The difficulty in externalising knowledge can be illustrated by reference to Popper's 'third-world knowledge' — he holds that words on paper in filing cabinets are knowledge (Magee, 1973; and Popper, 1963). But a filing cabinet cannot exercise knowledge. More importantly, the knowledge invoked in a person who later reads the words will not be the same as the knowledge that prompted the author to write the words. If we were to accept the words themselves as knowledge, we could never get to know what the words know.
16. This condition of linearity applies to both classical logic and its derivatives, for them to be computable, and this condition remains

necessary even in the case of parallel processing, which refers to simultaneous execution of processes along separate linear paths.
17. Putnam (1978) gives interesting insights into formal intuitionistic logic.
18. For a discussion of the contribution from computational techniques of AI to philosophical thought, see Sloman (1978), and for philosophical critiques of AI see Dreyfus (1979) and Searl (1984).

CHAPTER THREE

DESIGN AND THEORY

If I want to know what I am doing, then I need a separate description of my doing it, a theory: we need theories in order to be able to tell other things, computers, what we want them to do — and what we are doing is designing.

Turning from the more formal issues presented by computers, we will now look more closely at perceptions of design. This chapter will discuss design practices, with particular reference to the architecture of buildings. The motivation for doing so is the expectation that CAD systems can be useful to designers, and a belief that designers should use these systems. Design is seen as an application field for computers. As soon as we evoke this ambition for computers, we have to recognise that the criteria for a successful system are determined by the views designers have of their own world. Applications have to bridge the formal concerns of theoreticians and system developers, defined in detached and tidy worlds, and the practical concerns of designers responding to some whole and messy world. This distinction can also be put the other way round. We have to bridge the abstract world of researchers and the real world of practitioners.

By describing the practice of design we will be setting out the basis for a theory against which we can assess the performance of existing CAD systems, and which can be used to identify required characteristics of future systems. Theory means simply some abstract and generalised description of design that can correspond to practical instances of design. Abstractions entail the need to realise a theory in some form of expression and here I will do so quite informally in natural language. Theory also entails commitment as distinct from representational formalisms which are intended to serve as neutral enabling mechanisms. I will describe what I think design is.

We all know what buildings are, but what is it that we know and how does our knowing affect the demands that people make on designers, and the responsive behaviour of designers to those people? I will start with a discussion of buildings as design objects, the functional and aesthetic or expressive roles of these objects, and consequences for procedures that can be employed in design practices. Then I will generalise these observations to construct a theory of design. Lastly, I will offer a speculative sketch on design quality, which poses questions about the effect of technology on design expression.

DESIGN OF BUILDINGS

Designers design objects which are called into existence by other people. To understand design and to identify what may be particular to the activity of designing, we need to consider how these objects are viewed by other people. The artefacts designed by architects — buildings — exhibit properties which influence how buildings can be designed. Singly, these properties are not unique to architecture or buildings, but collectively they constitute a domain that differs from many other specialist fields.

Objects

The concept of a building is not well defined. We cannot answer the question: 'what is a building?' with a formal and precise definition that will distinguish any instance of a building from other artefacts that are not buildings. On the other hand, the concept of a building is well understood, and instances are recognised by convention. People nod their heads in agreement that certain objects are buildings and others are not. Change the people and their recognition of objects as buildings may also change. Thus, as illustrated in Figure 3.1, we are used to recognising certain buildings as houses,

Figure 3.1 *Things and houses*. Our recognition of houses as distinct from other things is dependent on shared conventions among people — change the people and different things will become houses. This dependence on conventions also applies to differences between classes of buildings — we do not now how we differentiate between classes. In general, we do not possess a definition of objects that are buildings, which differentiates buildings from other things.

Design and Theory

despite the differences between instances and our lack of a complete definition of a house.

Buildings are not solutions to problems in the sense of problems with defined solution paths, nor do we have criteria by which a single correct solution can be recognised. For any stated range of objectives which cause a building to be designed, there can be many different solutions. Assessment of any one solution will exercise people's judgement, calling on criteria invoked by experience of the solution which can be different from the criteria included in an initial statement of objectives. Good buildings are decided, never proved.

Buildings are not defined by function in the sense that their form is not determined exclusively by single (or few) well-defined utilitarian goals. A building 'has to stand up and keep the rain out', but it has to do much more. A building has to satisfy many loosely defined human demands, to fulfil non-material aesthetic needs. Use of buildings as expressive devices, as a means for passing messages between people, survives despite the reluctance of many achitects to acknowledge the importance of aesthetics. We need only to hear people describe a new community hall as looking like a factory, or new private sector housing as looking like a council estate, or see how much money a company will put into the image projected by its new office block to know that aesthetic sensitivity is still very much alive. Such manifestations are illustrated in Figures 3.2 and 3.3.

Buildings do not move. In general, buildings are assembled at their permanent locations. Buildings, therefore, are not subject to consideration of power-to-weight ratio, which means that they can carry functionally redundant material for aesthetic purposes, as illustrated in Figure 3.4. In this respect, buildings differ from many other designed artefacts such as aeroplanes and electric kettles.

Buildings are expensive. They require far more money than most people who use them can spend on anything else during the whole of their lives. This has the effect that decisions on what to build and how much to spend are delegated upwards to other people whose criteria might not be shared by 'everyone'. Buildings can become remote from the interests of the people they are meant to serve.

Buildings are experienced by everyone. You do not need to be a specialist to know about buildings. Anyone can pass valid judgement on whether a building is good or bad.

Demand for buildings is subject to violent and unpredicted fluctuations. Demand varies with people's changing perceptions of need, and need can be affected by people's redefinition of buildings. Housing provides a good example. During the period of general

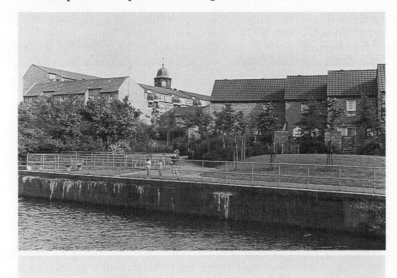

Figure 3.2 *Aesthetics alive*. People recognise the difference between public sector (above) and private sector (below) housing, and they do so by interpreting the visual appearance of what they see. People's expectations of difference have to be fulfilled by designers, by aesthetic expression. Aesthetics has deep roots, in this case in people's perception of ownership and status in a class divided society.

Scottish Special Housing Association, housing in Leith and Miller Homes in Barnton, Edinburgh.

Figure 3.3 *Appearance of money*. Organisations in the business of money evidently know the value of projecting confidence through visual appearance. They use aesthetics and are willing to pay for it. They pay far more than would be necessary purely to accommodate their material needs. Idealism operates equally where money counts, and where people profess altruism — the messages expressed may be different.

Scottish Widows' head office, Edinburgh.

Figure 3.4 *Just a house*? Buildings are not defined by function and, in general, they do not move. Tolerance of weight has enabled buildings to persist in an expressive role, passing messages between people. In this row of houses, what are these functionally redundant columns saying? Who felt so strongly about saying it to justify the investment of energy in realising this expression?

expansion in the 1960s, there was a widely held conviction that more buildings were required. It was believed that in the last quarter of this century we would have to build more new buildings than the total of all buildings built during the whole history of mankind. Need was manipulated by redefining slums or unacceptable existing housing stock, with effects as indicated in Figure 3.5, and enthusiasm was reflected in competition among successive governments over the numbers of new houses they built. By the end of the 1970s, during a long period of recession, we began to envisage a future in which no new buildings would be required. This was reflected in official arguments that claimed we had sufficient houses but that people were in the wrong place, leading to a shift in emphasis towards rehabilitation of old housing stock, as illustrated in Figure 3.6. The extreme manifestation of this shift was the official moratorium on all new public authority house building, in 1980. The interesting point is that the physical presence of artefacts which we recognise as houses and the circumstances of people who require houses did not change in any way commensurate with the change in demand for housing. This change in demand can be explained, but the explanation does not rest with designers of buildings.

Activity
Moving from building to architects, we can now explore implications of these design objects on architects' work-practices. The single most important point to emerge is that the work of architects can be, and is, judged by everyone. It is true that direct contact with architects may be limited to a few clients, or to client agencies who control investment in buildings, but the artefacts that architects design and on which they build their reputations can be experienced by anybody.

An architect's survival depends on his or her performance socially, as a person keeping in touch with the varying concerns of other people, as well as his or her special ability in translating these concerns into artefacts which we recognise as buildings. He or she cannot refer to an established body of expert knowledge in order to demonstrate that a product is a correct design. If other people declare that they do not like the proposed building, it is wrong. If the architect is sufficiently convinced of the value of the design, he or she can still make the wrong building right by exploring the concerns of these other people and convincing them that these concerns are accommodated in the design. But the architect will need to be receptive to any new information from critics and build up new arguments in order to be convincing. The received

Design and Theory

Figure 3.5 *New for old*. By defining slums we create need for new housing. Perceptions prompted by immediate social concerns motivate the design of the new buildings. But the process remains constrained by people's sense of aesthetics and results might also be viewed as slums.

Salford redevelopment in the 1960s (Bijl, 1968)

Figure 3.6 *Old into new*. An example of design which expresses sensitivities inherited from pre-war concepts of good urban planning (above): low-rise housing interspersed with tall blocks, plus large communal open spaces. These buildings have suffered from spontaneous rejection, with houses boarded up and empty. In this case, subsequent action has been to embark on an extensive face-lift (below) and to change title from public to private ownership. As well as improving functionality, by adding pitched roofs to keep the rain out, the change in appearance is evidently considered important to the revitalisation of this estate. Pilton redevelopment, Edinburgh in the 1980s

Design and Theory

information and what needs to be done with that information may change with every design project.

The necessary responsiveness and answerability conditions the kind of procedures architects can apply to information and, therefore, how architects are able to work. The relationship between answerability and procedures is illustrated in Figure 3.7, differentiated under learned disciplines and experience-based practices.

In a learned discipline people are answerable to their peers, to other similar people. These people can behave as though they share the same knowledge and, for this purpose, it is important that the knowledge exists in some externalised form. The knowledge is thought to have an existence detached from people. Relatively stable knowledge fields permit the evolution of formal procedures for operating on information, which are themselves recognised as valid. Execution of correct procedures contributes to the validity of products. Furthermore, the same stability that fosters such pro-

KNOWLEDGE BASED DISCIPLINE

ACTIVITY: explicit knowledge forms the primary means for progressing from problems to solutions

ANSWERABILITY: to peers who share a common knowledge base

DISCIPLINE: formal and detached — shared basis for assessing new developments

PROCEDURES: methodical and stable — overt exclusion of intuitive judgement

EXPERIENCE BASED PRACTICE

ACTIVITY: relies on explicit knowledge plus intuitive judgement of practitioners

ANSWERABILITY: to public who do not share the same knowledge base

DISCIPLINE: informal and inconsistent — practitioners must be responsive to a volatile world

PROCEDURES: idiosyncratic — loosely defined integration of knowledge and experience

Figure 3.7 *Answerability and procedures*. In all situations in which people are answerable to other dissimilar people, where participants do not share an established knowledge base, intuitive knowledge within persons is decisive in determining actions — intuitive knowledge here refers to that knowledge which is acquired through a person's experience and which cannot be externalised so that overt expressions alone substantiate the knowledge. The absence of shared knowledge among designers and other people leads to dependence on idiosyncratic design practices.

cedures also ensures that the procedures can be used repeatedly over substantial periods of time. Thus, we can accept in science a correct test of a hypothesis; in accountancy a correct balance sheet; or in law a correct legal transaction.

The difference presented by more informal practices, and by architecture in particular, rests on the question of answerability. Architects cannot call on the authority of a body of expert knowledge, or the design procedures it might engender, in order to justify their design proposals to other people. Recipients may not be interested in how a design was produced but they will still pass judgement on the result.

If architects and the work they do cannot rely on the correctness of overt and stable procedures, how do architects function? To find the answer, we have to look further at the link between procedures and knowledge. In doing so, we have to moderate the above differentiation. Disciplined fields are those in which we believe we have a shared body of overt knowledge which tells us sufficient to give us confidence in our further actions. Procedures then provide the means for moving between different states of information when we seek to apply our knowledge to achieve some stated goal. This connection between knowledge and actions is not exclusive. When members of a discipline are answerable to other people, either within or outside the discipline, we have to acknowledge that reliance on overt knowledge has to be supplemented with personal experience. This form of experience can be thought of simply as learning acquired through direct contact with our world, which remains within persons and which cannot be articulated overtly as shared knowledge. Because such learning is individual it will differ in different people, and differences might be manifest only implicitly in actions. We recognise such inner learning as intuition.

Even in highly disciplined fields an overt body of knowledge will never provide a complete explanation for everything that will be perceived in an event. This qualification applies both to events that are naturally occurring and events that are conditioned by the involvement of different people. We can say that the success of a person working in any field is dependent upon the state of established knowledge, the skill of the person in operating with that knowledge, and, perhaps most importantly, the person's use of intuition to supplement established knowledge. This position applies in science, the arts, and, especially, design.

We now come to innovation, the generation of something new, as a responsibility of architects. Innovation is more than the creation of a new action or product. Such newness is commonplace and

Design and Theory

ordinarily can be generated by overt knowledge or by accident. Innovation is made up of a sense of purpose, a route to a solution that is unexpected and unrepeatable, and a result that is surprising and subsequently recognised as valid. In this sense innovation occurs in all fields, including science. Einstein's hypothesis that light rays curve was innovative.[1] In architecture, however, the process of validation involves judgement and consensus leading to popular acceptance.

Innovation in architecture is important precisely because the definition and purpose of buildings is not well defined. Why people want buildings, what they expect buildings to be, and what they intend to achieve with buildings changes as people change. People's changing perceptions, needs, and social order lead to new demands. Only some demands will appear explicitly in briefs for new buildings, and some of these may be met through methodical application of knowledge. Other demands will be latent, becoming apparent only after a solution begins to materialise, and only then will they be recognised as necessary demands. The architect's role as innovator is to perceive unspoken demands, to give these demands expression in designs, and to translate these demands into physical buildings. Innovative buildings are unexpected and obviously 'right' in the perceptions of people for whom buildings are designed.

Procedures

What does this outline of design activity indicate for design procedures? Given that architects need to employ their own intuitions and, furthermore, that any established body of knowledge will be in a state of evolutionary change, demarcations between overt design procedures and intuitions will be imprecise and unstable. This does not mean that there are no tasks performed by architects that cannot also employ overt procedures. Clearly, at a basic level, architects use arithmetic rules for dealing with numeric data, grammatical rules for dealing with written information, and, perhaps, geometric rules for constructing drawings. These rules may be shared among all architects and other people. Architects may also share procedures for dealing with particular analytical and quantitative tasks on the periphery of design. But at these higher levels, where procedures have a more direct influence on outcomes, we become increasingly less certain of commonality. This uncertainty extends throughout the design process, from sketch design to production drawings.

In the course of designing, architects have to exercise control over demarcations between overt and intuitite design procedures. Any externally imposed and fixed demarcations will have the effect of reducing the scope of a designer's intuition and, therefore, will reduce design responsibility. In order to practise architects must refer to their intuitive knowledge about other people and this reliance on intuition leads to the conclusion that design procedures must necessarily be idiosyncratic to individual architects.

We now have a picture of architects as necessarily idiosyncratic and innovative, drawing on the combined intellectual resources of established knowledge and intuition. They have to be responsive to unspoken varied and changing demands from other people. How does this perception match up with the idea that architects should use computers? This question underlies the discussions of computer applications in the following chapters. The question is important not only for architects. It is relevant to all people whose work requires them to interact with other people and to the ordinary activities of most people.

OTHER DESIGN FIELDS

Some might argue that the view of architecture I have presented applies only to exceptional cases; that this view is not representative of the bulk of more mundane design practices. It might be argued that fields other than architecture are subject to overtly defined bounds and well-established goals which foster practices based on established design procedures. If a general anticipation of using computers leads to this kind of argument, then we are likely to argue design out of existence.

Designing is essentially a way of responding to some requirement where the expression of requirement does not describe what you want. Designing is about defining something that you do not already have, something that does not yet exist. If you know what you want and what you have to do to get it, for example, by employing some process of selection and addition applied to a stock of pre-existing things, then you are not designing.

Innovation is essential to design and distinguishes it from orthodox problem or puzzle solving exercises, and from routine executive or administrative processes. To convince ourselves on this point, consider the role of innovation in design as evident in the evolution of designed artefacts that have occurred over time, long periods of time, and consider how else they might have occurred.

Lastly, the extent to which innovation depends on intuition may vary in different design fields (Chapter 8). Intuition may be less dominant in those fields where functionality of objects plays a greater part in defining designed artefacts. But the role of intuition does not disappear even when designing aeroplanes or electric kettles.

Towards a Theory
With this view of design practice, we can now try to generalise our observations to construct a theory of design. We want to identify the conditions which need to be accommodated in the theory and which also will be relevant to any assessment of existing systems. We have seen that designers have to be responsive to other people. They have to talk to other people and their dialogue has to refer both to overt and intuitive knowledge. We can add that a designer's intuition must be allowed to modify overt knowledge as known by other people; otherwise there would be no need for designers. They must be able to express their contributions persuasively.

It can then be argued that the means of expression of designers, in the form of words and drawings, are employed in similar ways by all designers. These means are already dependent upon theory for purposes of interpretation — as is evident in the use of formal structures in expressions that have to be learned. This dependence on theory will increase if the activity of designers is to be supported by computer technology. Here we can conclude that theory should not be targeted at outcomes of design activity but at the expressive environments in which designers work, at the things that designers do in these environments.

Design generation
Design has to achieve a fusion between things to create new things, so that the products are recognised as having a right and proper place in the world of people. These things should be understood as referring to anything: physical objects, abstract ideas, aspirations. Things are extracted from some design context, transformed through design, and results are replaced. In the example of buildings, the context is people and results have to be assessed by reference to people.

This generative view of design differs sharply from the more orthodox analytical approach to design systems. In the latter case, the usual approach is to decompose a given problem into parts until separate parts are recognised as being amenable to known evaluative procedures, and results are then aggregated into a solution. This approach has a peripheral role in design, when evaluating selected aspects of tentative design proposals. However, the absence of overt and widely recognised criteria for design objects excludes design practices from the mainstream of analytical developments.

A criticism of the generative view is that it does not offer clear targets for techniques which we can think of reproducing in computers and which will ensure better design products. Techniques themselves cannot ensure better design products, but they may be used by designers so that they can make better products. The generative view implies that designers must have control of the use of techniques, including the possibility of redefining techniques, in the course of generating designs.

Problem solving

In moving towards a theory of design, we cannot expect to explain all that goes on within designers when they design particular things. However, a theory must explain enough about the whole design activity in order to define the intended relationship between the theory and design practices. By targeting theory at the expressive environment employed by designers, we still need to know something about the whole process. The following paragraphs point towards a holistic theory of design.

By focusing on design as something people do rather than on the artefacts which are the products of designing, we can develop the view of design indicated in Figure 3.7. This view recognises that design products are inherently not predictable by overt procedures alone, or by problem solving techniques operating within a computer. Design is more aptly characterised as an activity of problem definition and event exploration in which partial results lead to problem redefinition.

Now let us look a bit more closely at what is meant when we say that design is not problem solving. A typical problem solving approach can be identified by the following necessary constituents:
    (*a*) a known state of being, within a single well-defined domain;
    (*b*) knowledge of procedures available within the domain, by which a given state may be changed;
    (*c*) a goal, expressed in terms that:
        (*i*) specify some new state, including the criteria by which it will be recognised;
        (*ii*) specify boundaries to the range of procedures for changing the existing state.

When we add that a formal problem solver has to be defined solely on overt knowledge, then we will find that problem solvers exclude design. From previous discussion, design does not offer the necessary goal specifications.

Any problem solver that operates in a detached mode, indepen-

Design and Theory

dently of a person, is dependent on the following further conditions. Any instance of problem must have a start and a finish, and must be contained within those boundaries as a discrete and whole thing — typically, any problem and goal specification can be undone by moving those boundaries. The wholeness of a problem is then decomposed into discrete parts, as sub-problems, until parts present the conditions required by the available change procedures. To ease the task of matching parts to procedures, emphasis is placed on prior typing of parts. Sequences of change procedures provide solution paths. Solutions are found by aggregating results, and problems have to have single (or very few) solutions. A match between a solution and a goal can then be recognised as a correct result. This view of problem solving rests on concepts which are important to the internal operations of problem solving systems. These concepts can be identified as:

(*a*) wholeness of things;
(*b*) differentiation between whole and parts;
(*c*) discreteness of parts;
(*d*) prior typing of parts;
(*e*) correctness of results.

These concepts have become widely and firmly established in many fields. Their effects on design are evident in CAD systems. What we find is selective decomposition of loosely defined design practices into well-defined sub-tasks that are amenable to a problem solving approach. So we get computer programs that perform analytic functions to evaluate thermal performance and energy requirements of proposed designs for buildings. In most cases, these systems contribute nothing to our understanding of design synthesis, where synthesis has to reconcile disparate interests in a design object. We have to consider how the concepts that support problem solving relate to design.

(a) Wholeness:
In general, the start and finish of a design process appears to be circumstantial. The context in which design occurs is people and they decide when a design is to start and finish. There are few overt criteria for recognising that a design is complete — in the case of buildings there are none.

The boundaries to any instance of design are variable. A design process can include responses to ever more unforeseen considerations. In the case of buildings, these considerations refer to unbounded domains presented by any groupings of people (Figure 3.7). We do not know how to draw a boundary around some total design interest in a building so that we know we have the whole building.

For example, the design of a door handle sits in the context of a door, in a room, in a dwelling, in a building, in a locality, in a town. Equally, the door handle sits in the context of fire safety, affecting the ease of escape through doors by people with burnt hands and, therefore, it is relevant to social concern for welfare and health provision. And the handle has to be manufactured with available technology, marketed, and installed. We can say more about the door handle, but we do not know where to draw a boundary around it to define a complete and discrete design interest.

(b) Parts:

Decomposition of design into parts has the effect that the parts are differentiated according to interests from different domains, with no overtly known relationships between domains which can be used to compose results into whole design solutions. In the case of buildings, we do not know how to add the result of a thermal performance evaluation to a result of a daylight distribution appraisal.

Design has to reconcile arbitrarily different and contradictory interests in a design object, and this reconciliation cannot be achieved by overt processes alone. Parts are defined by the perceptions of different people, and synthesis with respect to any perception of whole is dependent on personal contributions from those people.

(c) Discreteness:

If we decide that parts do exist, we do not know how to define them as discrete parts. In the case of buildings, a part tends to be defined by the context made up of other parts, and changes to the context change the part. It is not practical to work with discrete parts where changes to one part are likely to propagate unforeseen changes to many other parts.

(d) Prior typing:

Parts occur as unforeseen instances which undermine confidence in prior typing of parts. Attempts at prior typing either have to be undone, as new instances make unforeseen demands on type definitions, or they compel conformity of instances which adds extraneous constraints on a design product.

Prior typing is critical to any formal theory intended to be implemented as a computer system which will perform tasks. Any expressions that describe a design object, from a designer, must

Design and Theory

have formal interpretations that will invoke functionalities within the system, and these interpretations rely on typing of parts of the designer's expressions.

(e) Correctness:
Lastly, we come to the concept of correctness of results. Here we have to recognise that design briefs generally do not include explicit specifications for goals that can be matched against design solutions. We cannot know that we have correct results in any calculable or logically provable sense. In the case of buildings, we can observe that there is no abstract formal definition of a building that will enable us to tell buildings apart from other things. Nor do we have rigorous classifications of buildings that differentiate between, say, houses and offices among all other instances of buildings. In the absence of such knowledge, goals cannot be explicit and we have to accept uncertainty in our solutions.

AMBITIONS FOR CAD

Having described design as *not* problem solving, how else can CAD perform useful tasks? By now, many researchers readily accept that computers cannot deal with the whole of design. The defensive position is to claim that there are parts of design which are computable, that these can be differentiated from other parts which probably will always remain the responsibility of human designers, and that these parts can cohabit in some whole perception of design. This presents a focused view of the position generally claimed for computers in many other fields of application.

It is argued that there are things that different designers do repeatedly in a similar manner. It should therefore be possible to identify regularities in the occurrences of these things, so that the regularities can form the basis for computerised representations. Examples are discussed in Chapters 4 to 6. Important to that ambition is the idea that sub-tasks executed by computers should be recognised as being useful by designers and other people involved with designs.

Three main problems stand in the way of this ambition. First, we have the problem of similarity. If we, as system designers, recognise similarity in occurrences of tasks, it does not follow that the people performing the tasks will see the same similarity. Our perception of tasks may be conditioned by an intention to represent them in a computer, which will cause us to see only those things that we believe we can represent and filter out all else. The designer who is performing a task, with no anticipation of it operating within a computer, is likely to have quite a different perception of what he or she is doing. This problem is evident in familiar demarcation disputes about quantitative and qualitative aspects of design tasks. If we require designers to use the results of tasks executed by computers, we pre-empt such disputes but we also impair their efficacy as designers.

Secondly, we have the problem of change. If we are able to satisfy ourselves that we can recognise tasks that do indeed recur in like manner, it does not follow that the same tasks will, or should, recur indefinitely. Experience of design shows that the same task, viewed after a time interval of years, is likely to exhibit a different formulation and execution — the task will have evolved. The act of changing may be imperceptible, having resulted from responses by many designers to a changing world of design. As long as it remains people who execute design tasks, they will be able to accommodate such evolutionary change. Computers do not (yet?) have this ability. Usually, a task set up in a computer has to go on being used until its results are judged to be wholly unacceptable, and then the whole of the task has to be discarded and replaced. In this way computers imply a revolutionary rather than evolutionary development of design practices. Working within such rigid anti-evolutionary systems, designers will not be able to design.

Thirdly, we have the problem of differentiation. If we can satisfy ourselves that we can differentiate between sub-tasks of design which are indeed computable and other sub-tasks which will remain the responsibility of human designers, it does not follow that computers will be able to execute their tasks. By recognising that some sub-task has to remain a human responsibility, we are admitting that we cannot describe the task. This has the further implication that we cannot describe the boundary to the task, which means that we cannot be confident about any interface between the task and another task within a computer. If we then expect designers to exercise their responsibility, we should expect that things said by designers to computers will deviate from the prior definitions of computerised tasks. A computer will not be able to recognise that it has to execute a task. Complementary and effective differentiation between computable and non-computable tasks appears to depend upon prescriptions for non-computable tasks. Again, working within such prescriptive systems, designers will not be able to design.

What, then, can we expect from computers? We can return to the definition of design as problem definition and exploration, and accept that computers are essentially problem-solvers. Computers are symbol processors and the tasks which they can be made to perform lie within the bounds of computational symbolic logic. The ambition is to enable designers to define their own problems, with respect to their perceptions of design domains, and redefine their problems in the light of responses from their computers. These responses should be determined only by the general logical operations of a system, invoked and controlled by designers, and these responses should support designers' formulations and evaluations during the course of designing. To progress in this direction, we will need representation environments which are meaningful to designers, in which models of design problems will have interpretations into low-level functionalities available from within computers. The key point is that a representation environment should have no in-built anticipation of design domains, but it must support concepts which will allow designers to formulate and reformulate expressions of their own design knowledge.

Designers will need to know the representation environment in order to exploit the logic that the system employs. Designers should also be allowed to use expressions which do not have interpretations in that logic but which designers and other people are able to interpret. Ideally, we want to differentiate between the structure of expressions accepted by the representation environment, which needs to be followed by designers, and the content of expressions, which can be decided by designers. Structure is needed to ensure effective interpretations from a designer's speculative problem definitions to a computer's logical operations. Content refers to interpretations of expressions, that might have meaning only within designers and other people, by which a designer is able to assess responses that come from a computer. This position is not very different from other more familiar representation environments, such as writing or drawing, except for more extensive use of automatically executed logic.

We have now declared a certain ambition for CAD which rests on a theory of design, and the theory focuses on the activity of designing. This is a holistic theory which acknowledges idiosyncratic combinations of overt and intuitive contributions from individual designers.

This ambition is profound and uniquely linked to CAD, focusing on the conjunction of people and computers. This focus does not rest on an assumption that computational logic potentially can embrace all behaviour exhibited by all people, as appears to be the position of many researchers in AI. Instead, the focus is on human logic controlling logical machines which can be used to support any expressions that people may wish to pass between people. This ambition may turn out to be more profound than that of AI, and it may prove to be more influential on people's general acceptance of computer technology.

The next three chapters will describe users' experience of different computer applications to design. The discussions will relate practical considerations to the deeper issues underlying CAD and those chapters will be followed by further discussion of possibilities informed by experience.

QUALITY

To conclude this chapter, I will sketch some speculative thoughts about design quality. Once more, I will concentrate on expressions from people, the means used to realise expressions, and what people read in expressions. So far I have said little about what makes people like designs, the quality of designs. Some understanding of quality is appropriate, since it is likely that people's perceptions of quality influence their appreciation of designers, with implications for design practices.

In the case of architecture, architects and other people are worried about the design quality of buildings, as is reflected in the present rather low standing of the architectural profession. People commonly face the prospect of a new building with dismay, especially in environments which contain many old buildings, such as the City of Edinburgh. The creation of a new building ought to be greeted with enthusiasm, as an opportunity for expressing the vigour and hopes of people, as a celebration of our time. Why does this not happen?

The reasons for antipathy towards new buildings are no doubt complex, reaching into matters which extend beyond architecture. I will suggest one possible and rather fundamental reason. If it evokes recognition from the reader, it may have some validity. What we are about to discuss is an aspect of mechanisation; the use of machines as extensions of people, and its bearing on the way in which we make objects which exhibit design quality. This discussion is relevant to the idea of computers serving as extensions of people. However, what follows refers to a world that predates computers.

Contact between minds

As a broad generalisation, we can view people's regard for buildings as a response to what they see of other people in buildings. They see expressions of other people's existence in the physical appearance of buildings. This is a restatement of my earlier point about the expressive or aesthetic role of buildings. In old buildings, worked on by many people, we sense the presence of those people and gain reassurance from such manifestations of our roots in history. If this idea seems fanciful, it is evidently shared by many people, as in the example of Americans seeking out evidence of their forebears in Europe.

The expressiveness of an old building comes not only from its master builder, the architect, who might have decided its main form. It comes also from all of those people who were party to its realisation: the craftsmen who worked the materials used in its construction. Their minds, filled with knowledge acquired as people among people in the world of their time, naturally and inevitably found expression in what they did with their hands. The products we now see reveal their minds and allow us to sense their presence.

This conjecture is not meant to be taken literally, in the sense that people now are consciously able to interpret what they see as the minds of identifiable people from the past. Nor is it meant to be taken mystically, as though people now can see actual ghosts of people from the past. Instead, what is being suggested is that people now do sense something in what they see, perhaps intuitively, that this sensory stimulus links them with past human existence, and that this connection is felt to be reassuring. Thus, we find that people appreciate old buildings.

Such interpretations of visual sensory experience can be regarded as the meaning of aesthetics. The essential role of aesthetic expressions is to establish connections between people which may convey messages of peace and stability, or oppression and revolution. It is this understanding of aesthetics that makes the visual impact of Gothic cathedrals and de Stijl houses (Figures 3.8 and 3.9) so powerful.[1]

Hands

But why, then, are modern buildings so often viewed with such dislike? Here I suggest that a significant part of the answer is to be found in the link between minds and hands, and in the removal of hands from the processes of designing and building. What makes this link so important is that it makes expressions from the whole mind, of both overt and intuitive knowledge, possible. Where expressions are formed into artefacts such as drawings, sculptures, and buildings, it is the direct link between mind and hands within persons which makes such holistic expressions possible, and it is these expressions that people appreciate in old buildings.

The thrust of this argument refers to the way in which a mind can

Figure 3.8 *An old aesthetic*. The open crown of St Giles is from the fifteenth century and it sits on a building started in the twelfth century (Gifford *et al.*, 1984) — a time when people performed astonishing feats of construction to express their religious beliefs. The building suffered under the Reformation. It was improved (outside) at the time of the Enlightenment and was remodelled (inside) by the Victorians — improvements which illustrate the human syndrome of 'always knowing it better *now*'. We can read the concrete presence of the building as a complex aesthetic expression from the lives of very many people.   St Giles, High Street, Edinburgh.

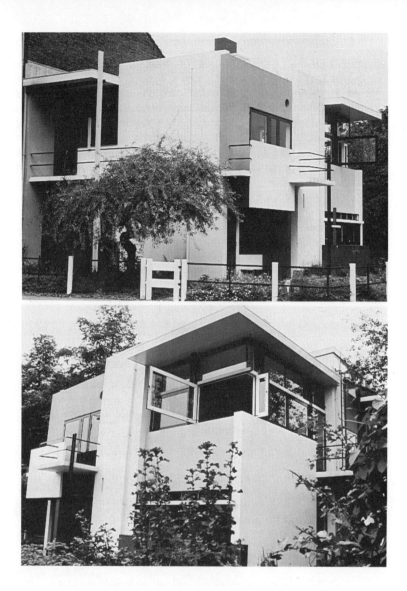

Figure 3.9 *A modern aesthetic*. Here we have a sharp-edged, machine-processed building of 1924, in contrast to Figure 3.8. Its expression is forceful as part of a cultural movement which embraced painting, literature, and music. It is a vigorous realisation of a machine-age aesthetic which spread into the homes of ordinary people. To succeed, designers and other people had to be active in a shared enthusiasm, to express their celebration of beliefs and aspirations. De Stijl can be viewed as a new broad-based aesthetic movement of our machine age and, significantly, it occurred near the start of this age.   Architect: Rietveldt, Netherlands.

control external manifestations of what it contains. For our purposes it is not necessary to know what is actually happening within the mind, we do not have to look inside people's heads. It is enough to observe its externalised manifestations and know that not all that is manifest can be explained or replicated by employing overt knowledge alone. Hands, as instruments for producing such manifestations, are not directed by overt knowledge and can be responsive to a person's intuition. Consider the examples shown in Figures 3.10–3.12.

Mechanisation
Developments in building technology have led to mechanised processing of materials, reducing the need for people's hands. This trend has become so pronounced that any act of removing hands is now regarded as a welcome advance. For every pair of hands removed from the job of making a product, the contribution of a mind is lost and, with it, an opportunity for holistic expression.

Figure 3.10 *A little piece of mind*. An ordinary bit of construction. The adzed joist (middle) shows the maker's hand in the surface pattern along its length, and the anchor in his mind is seen in the apparently spontaneous bit of carving. By contrast, the machined joist (nearest) shows only that it was passed through a machine, with no visible thought between start and finish.

The roof construction outside the author's office, Edinburgh.

Design and Theory

Figure 3.11 *Mind Over Matter*. A chest of drawers has a form but is not defined by function alone. Here the maker thinks about the artefact, speculates about the integration of its parts, and lets his mind pour through his hands to achieve a sensuous expression in wood. The mind of the maker then remains visibly present in the physical object.

Furniture maker: Chris Holmes, Edinburgh.

Figure 3.12 *Mind expressed through hands*. In the making of a violin, the maker has to decide on the composition of properties of the wood — its grain, density, thickness, curvature, and edge fixing — all combining to effect sound quality. Decisions pass through the maker's fingers, some deliberately so, and others more spontaneously from intuition. The maker's mind then finds expression in the individuality of each product, each an instance of the same thing: a violin.

Violin maker: Bert Houniet, Utrecht, Netherlands.

When a person is framing up instructions for a mechanised process two things can happen. First, the instructions have to be capable of being executed, so the person is constrained by knowledge of what the machine can do. The person's intuition is constrained by the formality of the machine's procedures. Secondly, the instructions may be targeted at the mind of another person who will operate the machine. As a machine operator, the mind of the second person is likely to be devoted entirely to the procedures of the machine, as an extension of the machine. This second person will have very little opportunity to give expression to human concerns that are of no concern to the machine. The result, unavoidably, is a product that is sparse in human expression.

What people see when they look at new buildings is the control of mechanised processes by a few individuals; by architects together with other professionals and their clients, very few people among many. Even these people are constrained by the existence of the processes, being reduced to selecting fully shaped products from pre-existing processes. People no longer sense other people in the fabric of buildings. They feel alienated from new buildings.

Here we have considered the dual effect of mechanisation in first enhancing and then restricting human abilities.[3] It appears that the idea that machines serve as extensions of people leads to the converse: that people become extensions of machines. I have emphasised the link between mechanised processes and the ability to manifest expressions from the mind. If we accept that design quality refers to people's interpretations of visual sensory experience and that satisfying interpretations are those which reinforce connections between people, then we should expect that any limiting effect of mechanised processes on people's ability to express themselves will lead to alienation. People will feel alienated from the quality of design objects, from buildings. At the start of this sketch I said that these thoughts would be speculative and they are intended to provoke further thought.

SUMMARY

A view of design practice, including designers, design objects, and the people who experience these objects has been presented. In the example of architecture, it was seen that design objects are not defined primarily by function. They also serve an expressive aesthetic purpose. They are costly, and anyone can pass valid judgement on whether a building is good or bad. The loosely defined nature of these objects makes designing an activity of problem definition and exploration.

Answerability of architects to other people means that architects have to employ the full range of their human sensitivities. They have to employ their intuitions and be innovative. Consequently, the procedures that constitute design practices have to be idiosyncratic to designers. This perception of design is not unique to architecture and it highlights essential aspects of design which need to be accommodated by any systematic treatments of the design process.

A holistic theory of design has been outlined which accepts that designers have to control their own combinations of overt and intuitive knowledge in the course of designing. This theory says that design objects are inherently not predictable by formal processes alone, but that designers use formalisms, as in words and pictures, to express design intentions to themselves and to other people. The theory is targeted at the expressive environments in which designers work and at the things that designers do in these environments.

Ideally, in any implementation within a computer we want to differentiate between the structure of expressions accepted by a system, which has to be followed by designers, and the content of expressions decided by designers. Structure has to provide access from a designer's speculative problem definitions to a computer's logical operations. Content refers to people's readings of expressions from outside the system. We envisage people controlling logical machines which can be used to support any expressions people may wish to pass between people. This is a major ambition and we should expect it to be the focus of research over many years.

Notes to Chapter 3
1. The example of Einstein's scientific innovation is nicely described by Bernstein (1973).
2. For an illuminating discussion on the meaning of architecture, see Norberg-Schulz (1980).
3. This discussion of mechanisation accords with Weizenbaum's (1976) description of the dual effect of tools enhancing and restricting human powers, which he discusses in the context of computer science.

CHAPTER FOUR

INTEGRATED DESIGN SYSTEMS

If I am to use a computer and it is going to do things I would otherwise do, how can I know what it will do, and what will it expect of me?: we will now look at some users' experience of CAD systems.

CAD applications first began to emerge in the 1960s. Different CAD applications have come and gone over the past two decades. So far, CAD is still not widely accepted in the architectural profession, for reasons that have been indicated in the last chapter. In other design fields, such as electrical and mechanical engineering, the use of computers has become more established. In those fields, the methods and goals of design appear to be more overtly defined, and they tend to remain durable for long enough to permit the development and use of computer applications. But even in those fields CAD contributes little to the central task of design synthesis, especially when the goal is to produce innovative designs. In this and the next two chapters we will look at experience in the field of architecture in order to highlight practical implications of using computers.

CAD applications can be classified broadly into three categories, roughly in ascending order of ambition. First, we have computers which store and manipulate words and drawings, and which do so without knowing what the words signify or the drawings depict. Secondly, we have function-orientated systems, which are developed to perform specific tasks where the programmed function has to correspond to a user's perception of task in the user's application domain. Thirdly, we have integrated data-orientated systems, which are targeted at whole design domains, which support a range of tasks, and in which the system is responsible for satisfying the data requirements of each task. We will look at experience of these three categories of CAD in reverse order.

Contrary to what might be expected of an orderly evolution of CAD, the first integrated design systems coincided with the first function-orientated systems and predate the function-orientated systems that are now being offered as commercial products. One might have expected a progression from the more tractable function-orientated approach, through accumulation of experience, towards the more ambitious data-orientated systems. However, these turned out to be separate and parallel streams of development, employing different and incompatible starting assumptions. The function-orientated approach will be discussed further in Chapter 5.

This chapter will start by outlining two integrated design systems, indicating the models of design on which they were based. The focus will then move on to the users of one of these systems, to identify the characteristics of the user organisation which were necessary to its adoption of CAD, to identify its expectations, and to describe users' experience in coming to terms with the system. Lastly, some lessons will be drawn from this experience, for new advances in technology and for new users.

INTEGRATED SYSTEMS

The history of the integrated systems approach reaches right back to the start of CAD in the 1960s. The initial assumptions that prompted this approach were that computers would be used by designers within design offices, and that computers would be used to store and operate on information that comes within the normal scope of design practices. With this beginning, ambitions soon went beyond the scope of available technology, but the goal of integrated systems still survives.

An integrated design system is one which employs a single model to accommodate all information describing a design object. The model has to correspond to knowledge supplied by different people throughout the design process. The system then has to be capable of supporting a range of operations on the same model and to advance people's different interests during the course of designing the object. In addition, the system has to be capable of interpreting information from people's overt expressions of their thoughts about a design object, from drawings and text, to supply that information to the model. This goal focuses attention on the possible ways of organising data within computers, a data-orientated approach. This characterisation applies to the technologies of the 1960s and 1970s; in the 1980s the same ambition is called a knowledge-based approach.

The general model of design implied by integrated systems may be summarised as:

Integrated Design Systems

(a) a design is a single coherent description (of a building) which can supply information for many varied design tasks;
(b) any part of a description may be defined by any other parts;
(c) any part may serve more than one task.

To this definition, we should now add that these systems must not rely on some prior specification of particular interests which people may have in design objects. These systems need to be capable of representing unanticipated design objects and they must be able to accommodate designers' own formulations of design tasks. These qualifications are necessary if we accept the discussion of design in the previous chapter. When early attempts were made to develop integrated design systems, we did not know how ambitious we were.

Two early examples

There are two notable examples of integrated design systems which were started in the 1960s and which were developed and used in practice throughout the 1970s. One is OXSYS, for designing hospital and ancillary buildings; and the other is the SSHA system, for housing design.[1] These examples remain instructive because of the range of detail information they covered, and because they were used in professional design offices. The experience provided by these systems, in terms of the extent of coverage of users' design practices, has not yet been repeated.

We will look at both systems to uncover the perception of design that each represented — their models of design. We will then look further at the SSHA system to consider users' experience of that system.

Both OXSYS and the SSHA system were similar in employing graphics to depict spatial properties of design objects, buildings, and in using text expressions to describe other properties. Each system's model could then represent detailed descriptions of materials and other properties associated with graphical depictions of buildings. The goal was to enable designers to draw general arrangements of building designs and use the system to generate performance and production information, as indicated in Figures 4.1—4.3. The principal difference between these two systems was evident in their decomposition of design objects. OXSYS decomposed buildings into parts which corresponded to discrete building components; and the SSHA system saw parts as corresponding to junctions which, in turn, were used to describe components.

Both systems were restricted to paraxial geometry. Lines had to be straight and parallel to the axes of a screen co-ordinate system, and they could meet each other only end-on or at right angles. This restriction applied only to that part of the graphics which supplied

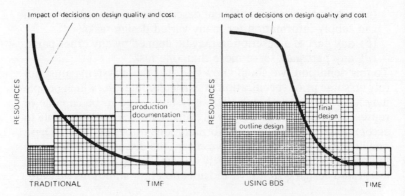

Figure 4.1 *The goal of integrated CAD systems*. The two diagrams show the relationship between design decisions and production information, in terms of effort invested in different stages of design. In traditional practice, least effort goes into early design decisions which have most effect on the outcome and performance of a building. Most effort goes into preparation of production documentation necessary to construct the building, but this has little effect on whether we want the building after it is constructed. The goal of early integrated design systems was to redistribute available effort by reducing the time spent by designers on preparing production documentation, and increasing the effort available for early design decisions. By freeing up the time available for design, it was believed that we would get better designs.

Figure by permission of Applied Research of Cambridge.

information to the system's model. Other 'cosmetic' lines which appeared only as parts of drawings could be arbitrarily angled or curved to depict such things as furniture in buildings or the slope of pitched roofs.

The limitation of paraxial geometry can be traced to the close link between drawing operations, knowledge of geometric entities, and the transformations that can be applied to them. Prior knowledge of geometry had to be encapsulated in program code so that the drawing operations that they supported could be available to a user. The user could not subsequently redefine geometry when using the system. At the time, paraxial geometry was as much as we could handle. Even now a truly general three-dimensional planar or curved surface geometry embedded in a building modelling system presents severe problems.

OXSYS model
OXSYS employed a concept of parts as physically bounded components which were fully described by reference to paraxial,

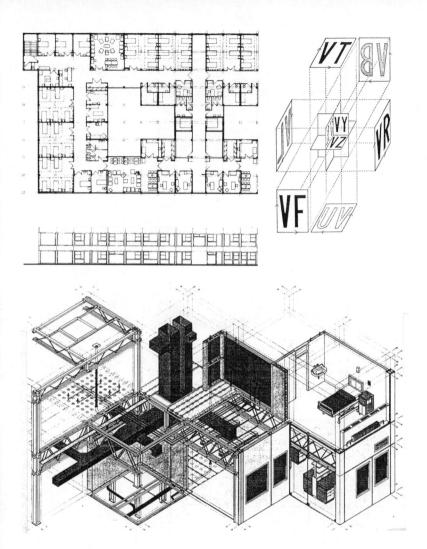

Figure 4.2 *The OXSYS system*. This system employed a concept of paraxial boxes (top right) which was used to describe components which could be assembled into whole buildings. Drawings of the shape properties of components were associated with the faces of (and sections through) boxes. Other material or functional properties were then linked to the same boxes. By assembling boxes detailed drawings of buildings could be obtained (top left) — note that the box structure is not visible in assembled drawings of components. From general arrangement drawings, the system was able to generate detailed production documentation. By employing this attractively simple concept the system was able to represent complex buildings in accordance with the Oxford Method of construction (bottom).

Figure by permission of Applied Research of Cambridge

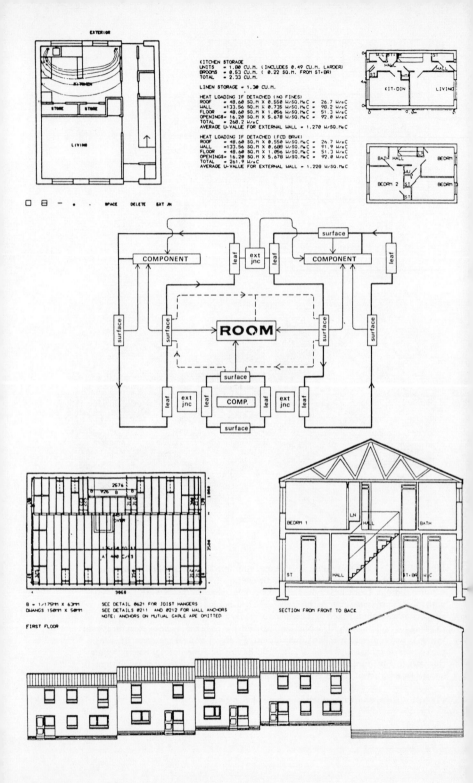

six-sided rectangular 'boxes', as indicated in Figure 4.2. The planar boundaries of the boxes did not have to coincide with the faces of objects — these were notional boxes used by the system. Boxes could be stacked alongside and on top of each other to produce assemblies of objects corresponding to compositions of parts to represent whole buildings.

The OXSYS model saw parts as named graphic entities plus associated non-graphic attributes and values. Note that we are discussing a model and its constituents as an abstract formal structure on which the system is built, and parts here refer to things which occur in the design object represented in the model. The model's entities are things inherent to the representation system onto which parts of the design object can be mapped. Attributes associated with an entity can receive values that match the values of corresponding properties of a part as known in the user's domain.

The shape attribute of any graphic entity was seen as a paraxial box, usually with fixed size values. Each face of a box could receive as an attribute a picture whose value could be any arbitrary arrangement of lines. The picture was arbitrary in the sense that the model gained no information from it. Typically, the lines would depict a user's view of the part of the building described within the box. Thus, an assembly of boxes could be depicted on a drawing as an assembly of pictures on the faces of boxes. An interesting achievement of this approach to modelling was that the resultant drawings could depict buildings without revealing the box structure of the models that were used to represent the buildings.

Non-geometric attributes associated with geometric entities, boxes, could refer to other properties of parts, such as material content and construction details determining permissible adjacencies for boxes with respect to other boxes. Relationships between parts were represented as attribute values within boxes.

Using this model, the system enabled a designer to describe a building by selecting and arranging component parts on a drawing as plan views of boxes on a computer display screen. Using computer operations programmed within the system, it could then

---

Figure 4.3 *The SSHA system.* This system employed three types of entity, plus relationships; junctions, components, and room spaces. Junctions defined components which, in turn, defined room spaces, and each could propagate changes to properties of others — a non-discrete component representation (middle). From materials and associated height values assigned to components, the system could generate section and elevation drawings and generate layouts for timber structures (bottom), and perform daylighting and heating evaluations (top), as well as supply quantity data for QS bills of quantities.

assemble different views of the building, providing a range of information. This included information on accommodation zones and schedules, lists of construction material quantities, thermal performance and services, as well as detailed production drawings. The designer was able to modify the drawing of a description of a building and the system could reveal consequential changes to this other information.

What were the limitations imposed by this model? General limitations due to software technology will be discussed in the example of the SSHA system. The most critical consequence of the OXSYS model was the manner in which it required information about the user's domain to be presented to the system before it could function. The OXSYS model required prior definition of items in component libraries, which had to contain size and other properties of all components (sometimes with pre-defined degrees of freedom), in detail, and in conformity with the entities built into the representation system. These library items had to be complete before a designer could identify and locate instances of components in the description of a building. The designer was not able to change a component arbitrarily in the context of other components in the course of designing a particular building.

The model rested on the concept of discrete components. Relationships between instances of components could not be expressed so that combinations of components could propagate consequential changes to other components.

The separate status of graphic entities, for the purpose of describing a building, meant that descriptions could be modified only by selecting differently named graphic entities. The shapes of parts of a building, depicted graphically, could not be changed by reference to other properties of the building. Moreover, junctions between components could be described only by reference to paraxial planes, thus limiting the system's knowledge of buildings to orthogonal geometry.

The view of design implicit in the OXSYS model was that:

(a) buildings consist of aggregations of discrete, manufactured components;

(b) information contained within descriptions of components remains relatively stable and is re-used on different design projects without change;

(c) arrangements of components are constrained to orthogonal geometry.

It should be stressed that these limitations were not unique to OXSYS and persist in other systems being developed today. The

Integrated Design Systems

use of notional boxes by OXSYS proved surprisingly powerful when dealing with descriptions of large and complex buildings, and the achievement of OXSYS is that it was conceived twenty years ago. Even more significantly, this integrated design system was used productively on large design projects over many years by the Oxford Region Health Authority.

SSHA model

The SSHA system was based on a different approach to components. Instead of one box per component, several boxes or stretchable 'thick slabs' could be combined to describe one component and all instances of components were described *in situ*, in the context of other components in the design of any particular building. A component received a general specification of material content, excluding boundary conditions, and the system looked at geometric relationships with adjacent components to identify necessary junction details (Figure 4.4) to complete each component description.

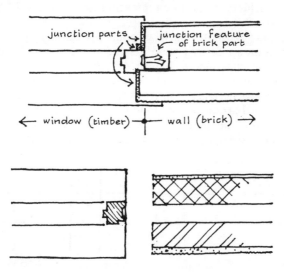

Figure 4.4 *Junction detail*. A typical junction between wall and window (top), in which junction parts modify descriptions of component materials to either side of the junction (bottom). In the SSHA system, a description of a junction detail included add and deduct quantities for the adjoining component materials, as well as descriptions of its own parts. Junctions were used to describe component instances.

Computer Discipline and Design Practice

This system saw buildings as arrangements of junction details which, in turn, described components filling the spaces between junctions. The faces of components which were not junctions were recognised as surfaces to room spaces. Rooms were themselves recognised as entities defined by components. Essentially, the model consisted of three types of entity: junctions which corresponded to construction details; components which corresponded to construction components; and spaces which corresponded to room spaces.

Junction entities were recognised by geometric configurations (Figure 4.5) which linked specifications of component materials to a library of construction details. From the library a junction received its description in terms of detail section and elevation drawings (Figure 4.6) as well as material unit quantities. Component entities were described by a name (wall, partition, etc.), plus attributes: primary material, internal cladding, external cladding (for external components). Component entities were related to junctions through leaves (parts of component boundaries corresponding to parts of their hierarchical tree representation, Figure 4.7). Spaces were described by a name (living room, kitchen, etc.) indicating function, plus attributes such as boundary surface finishes and furniture content. Spaces were related to components through surfaces. The diagram in Figure 4.3 shows that this model's structure was quite complex. This complexity was considered to be necessary to give designers control over the design of components in the tight spatial arrangements found in designs for houses.

In operation, the user drew arrangements of components to depict building designs, using stretchable graphic symbols to depict

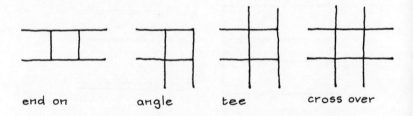

Figure 4.5 *Junction configurations*. Four types of junction configuration were permitted. Descriptions of junction details within the system's library included permutations of different component materials joined in these configurations. The system then identified particular configurations in drawings of general arrangements of components to identify the required details which would complete the descriptions of component instances.

Integrated Design Systems

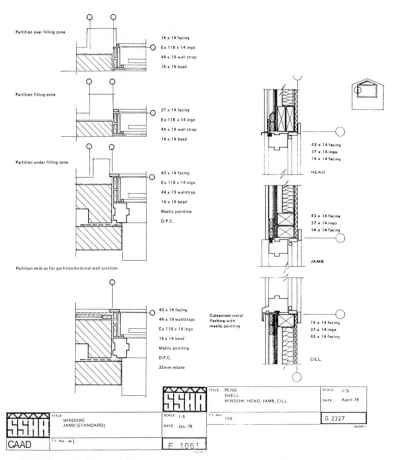

Figure 4.6 *Library of construction details*. Construction details were described by sets of detail drawings and their corresponding modules of materials quantities, which were prepared for different combinations of component materials in any or all of the four permissible junction configurations. The same details could be combined in different ways to define an extensive range of building components. During the design of a building, the system would recognise junction configurations from the plan drawing and identify corresponding detail drawings and materials quantities to complete the descriptions of components in the plan.

Figure by permission of the Scottish Special Housing Association

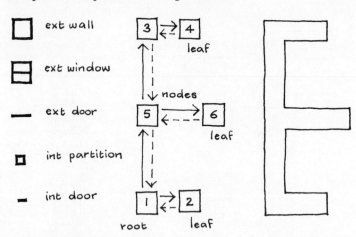

Figure 4.7 *Drawing symbols*. Five different symbols were used to draw arrangements of the fabric of buildings (left). Each filled a 300 mm or 100 mm square, corresponding to the grid squares in the drawing space on the display screen. Symbols were then planted on the drawing space to denote the start and end of branches of components, as seen in plan view (middle). These were held as logical tree representations in the system's data structure. The system would then join up the symbols to produce outlines of the components (right) — it drew a distinction between branches with the same materials specification forming internal junctions, and branches with different specifications forming external junctions. By searching its tree representation of components, the system was able to identify junction configurations and thus proceed to fill out the detail information for design drawings.

components (Figures 4.7 and 4.8). The system then identified the required construction details by recognising geometric configurations of components and relating these to each general specification of materials assigned to each component. The construction details selected by the system would, in turn, modify the materials quantities of adjoining components. To realise this system at the time when it was conceived, the range of possible junction configurations had to be limited to those shown in Figure 4.5, so that junctions between components would occur only along paraxial planes. However, by describing instances of components *in situ*, changes to a component could propagate consequential changes to adjoining components. This system did not rely on discreteness of construction components.

As an extension of this non-discrete approach, room spaces were

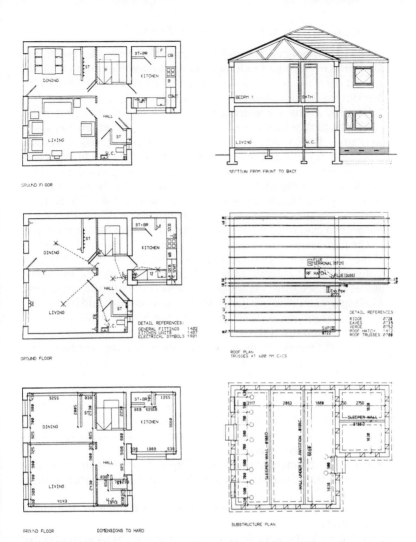

Figure 4.8 *Plan drawings of buildings*. A designer could describe a proposed building plan on the display screen, using the five symbols to position external walls, internal partitions, doors, and windows. While doing so, he or she could assign general materials specifications to components and label room spaces. The designer could then arrange furniture and fittings, choosing items from appropriate menus associated with room labels. At any time, the designer might make modifications to the plan, and could call for daylighting, heating, or structural evaluations. Plumbing and electrical services could be added, and the drawing could be dimensioned semi-automatically. Doing these things was simple, provided the designer remained within the boundaries of the system.

Figure by permission of the Scottish Special Housing Association.

linked to components through surfaces which served as 'junctions'. The surfaces to some or all components bounding a room could be altered from the room, in turn propagating further changes to the descriptions of the boundary components.

Graphic entities consisted of a small set of five stretchable symbols and these were used to position component junctions on a computer display screen. The resulting drawing was a system-generated depiction of plan views of rectangular slabs filling spaces between junctions. Users would read the drawing as a depiction of construction components which constituted the fabric enclosing room spaces. Height information was associated with material content, assigned by the user to the components. Thus, the system was able to generate elevation and section views of building designs.

The limitations inherent to the SSHA model were similar to those of the OXSYS model with respect to orthogonality and the separate status of graphic entities. Instead of a component library, the SSHA model required prior definition of a library of construction details. This library could be smaller, since the same detail could be used for many different components. The most fundamental difference was the use of non-discrete components, made possible by the SSHA model's inclusion of more relationships linking entities. This advantage was gained at the expense of more complex implementation problems, reflected in the difference between the relatively small-scale buildings that could be processed by the SSHA system compared with the large-scale buildings processed by OXSYS.

The view of design implicit in the SSHA model was that:
(*a*) buildings consisted of compositions of non-discrete components shaped or poured *in situ*;
(*b*) information contained in descriptions of construction details remains stable and is re-used on different design projects without change;
(*c*) arrangements of components are constrained to orthogonal geometry.

Like OXSYS, the notable achievement of the SSHA system is that it was conceived twenty years ago, and it was used productively on housing design projects for more than ten years by the Scottish Special Housing Association.

EXAMPLE OF A USER ORGANISATION
Now we will move our attention from these systems to their users. Before considering the impact of the SSHA system on the work-practices of its users, and their experience in coming to terms with the system, we first need to know something about the

organisation which sponsored its development and then used the system.

The Scottish Special Housing Association is a national body which operates over the whole of Scotland, usually on behalf of Local Housing Authorities. It has its own property development, professional services, management and maintenance departments, including architects, quantity surveyors, and engineers. It also has its own Building Department, which used to construct approximately half its houses. During the 1970s the Association built about 3000 dwellings per year, and it owned some 100 000 dwellings let mostly to Local Authority nominated tenants. This combination of responsibilities was and still is unique in Britain.[2]

Approximately half the Association's buildings take the form of two-storey terrace housing. The remainder includes flats, maisonettes, and other housing forms. Most buildings are of traditional brick or 'no-fines' poured concrete construction and, latterly, timber construction has also been used. In the past, most developments were undertaken on 'green field' sites. However, in recent years the emphasis has shifted towards infill developments to existing built-up sites, and to the modernisation of old housing stock.

Regularity of practices

Having a relatively stable workload over many years, the Association was able to concentrate a part of its resources on regularising its design practices. Responsibility for management and maintenance of its building stock motivated the Association to compile information on reliable practices as libraries of dwelling plans and construction details. This information is maintained and developed by the Association's Central Library, which serves separate project design groups. Extensive procedures exist whereby changes to the information, initiated either by project teams or by Central Library staff, are vetted and adopted as part of the Association's practices. These vetting procedures call on the experience of all the professional and other staff within the Association.

One consequence of these vetting procedures is that information needs to exist in a form which can be circulated to many different people and which is appropriate to their different interests. This led to the emergence, over some twenty years before the adoption of CAD, of an explicit procedural framework which was capable of handling information on construction details and dwelling plans. The resulting procedures for representing information thus became more durable than the instances of information contained in descriptions of particular building designs and construction details. The existence of this procedural framework presented a particularly favourable context for the development of a CAD system, and this framework made it possible to develop a practical system for traditional forms of building construction.

Since the introduction of CAD, the Association's practices have continued to evolve. During the 1970s procedures had to become still more rigorous in response to experience of machine operations, but, paradoxically, the building designs flowing from these procedures could be more varied. Improved access to information and the ease of preparing production documentation were perceived as making it possible to apply efficient practices of one-off designs. During the 1980s the context and the objectives of the SSHA have been affected by reduced national emphasis on building new houses. The role of the SSHA has changed, resulting in a concentration of its remaining resources on the modernisation of old housing stock. In effect, its brief for CAD changed.

Specification for CAD

The SSHA system consisted of two distinct parts. One dealt with the superstructure of houses and only with information describing superstructure between end (gable/mutual) walls and from the ground floor level upwards. This part employed the model which has already been outlined. The second part dealt with housing site layout and external works, covering all variables affected by placing houses in any spatial arrangement on any particular site. This part included the junctions between houses and junctions of houses to the ground. Briefly, the model for housing site layout employed polygon entities and set operations on polygons to represent sites as wholly covered by contiguous houses, garages, parking, roads, footpaths, gardens, and landscape areas.[3] The site layout system also included ground height modelling to fit site layout elements to ground forms.

The purpose of the housing superstructures system was to receive architects' design drawings, and generate production drawings and construction quantities. Further tasks for the system included analysis of space occupancy, daylight distribution, heating requirements, and the design of steel reinforcement for concrete and timber construction for floors and partitions.

The purpose of the site layout system was to generate information on vertical alignment of houses, roads, and other site elements, and provide production information for junctions between houses at mutual walls, between houses and the ground as underbuilding, and

Integrated Design Systems

between other site elements such as mutual edges between footpaths and landscape areas.

Computer resources
When CAD was first introduced, the Association already had experience of computer applications in its Quantity Surveying Department. These applications used external bureau facilities, and the Association had no experience of employing its own technical computer staff. The Association's policy was not to acquire its own computer. However, within a period of four years, the Association first acquired its own minicomputer to serve as a graphics satellite linked to a bureau computer, and then it acquired its own in-house central computer (a DEC System 10), plus a number of satellite minicomputers. Its central installation grew to provide a highly centralised service for most of the Association's departments, including the CAD activities of its Central Library.

Implementation of the SSHA system coincided with this rapid introduction of other major computer operations within the Association. During this period the Association had to come to terms with employing a new kind of person — computer systems staff. It had to assimilate new and unfamiliar knowledge. The management of the activities of computer staff by senior people who feel vulnerable on technical aspects of computing inevitably presents difficulties. It took time to create appropriate new employment conditions which would reconcile the different interests of people in user departments and in the centralised systems group. In parallel with this period of adjustment to computers in general, it took more than four years to establish a CAD development group and implement the SSHA system.

Benefits from CAD
The Scottish Special Housing Association invested several hundred thousand pounds in CAD. This was a great deal of money in the late 1960s and early 1970s. The scale of investment was partly due to the Association's pioneering entry into this field. The Association made the judgement that its goals were practical and the scale of its operations meant that it could bear the cost. It believed that results would be financially beneficial. In retrospect, it is possible to identify substantial benefits, amounting to millions of pounds, experienced by the Association and its client community.

The primary objective for the SSHA system was to increase the speed and ease of obtaining production documentation. Despite the unusually explicit organisation of SSHA practices, and despite the use of common house plans for different projects, it was said that more than half the available design time of SSHA staff was being spent on modifying existing production drawings for re-use on new projects — scratching tracing paper with razor blades. Modifications were largely the consequence of varying interpretations of regulations and by-laws in the different Housing Authorities where projects were built. It was then argued that design quality is dependent upon the skills and experience of individual designers, and that improvements in design quality should follow as a consequence of reducing the time spent by those persons on preparing production documentation. This argument, subject to minor variations, has become an orthodox justification for CAD in general.

The SSHA system employed its model of buildings to develop and analyse proposals presented by designers, permitting arbitrary changes during the course of design, and the system automatically generated very detailed building production information. Benefits were seen in terms of increased speed. This led to earlier building starts, thereby reducing the effects of rising building costs. Increased speed also permitted preparation of alternative production documentation to take account of local fluctuations in availability of labour and materials, resulting in further cost savings. The information available from the system also made it possible to exercise more control over expenditure during construction, thus allowing building starts in cases where tendered prices were precariously close to cost targets. Such benefits, in combination, have resulted in dwellings which would otherwise not have been built being available to house people.

To a significant degree, this success can be attributed to the system's ability to interpret drawings in terms of other nongraphical information, in other representations of knowledge about the objects depicted in drawings. It was the system's ability to invoke and operate upon expressions in the form of drawings and words which was useful to the Association in its dealings with client bodies and the construction market. This ability was instrumental in realising cost savings. It should also be noted that it was the regularity of the Association's design practices that made machine interpretation of drawings possible.

Experience of Users

We will now consider the experience of people working in a part of the user organisation whose work is not obviously design but is closely associated with design. We will focus on the work of quantity

Integrated Design Systems

surveyors and on their understanding and acceptance of CAD. In doing so, we are selecting the part of the organisation that appears most amenable to being helped by computers, but, since quantity surveyors are at least partly answerable to architects, we should expect some difficulties. If the experience of quantity surveyors within this design organisation poses serious questions about the use of computers, then we may expect that the same questions, with greater force, would also be significant for architects.

We will be referring to the experience of an actual user organisation. Although some of this experience may appear odd, suggesting that sensible people could have done it better, it needs to be stressed that this experience refers to a highly sophisticated and well-organised body of people. Far from criticising these people, lessons from this experience tell us more about our inadequacies in formulating systems.

The particular aspect of the SSHA system under discussion, that of quantifying and costing designs formulated by architects, was not central to the problems involved in developing this system. But this aspect was seen by the project sponsor as the primary justification for the system. The goal, as seen by the sponsor, was computerised quantification of architects' drawings through the intermediary of interactive computer graphics. This goal differed from the more usual practice of processing quantities which are manually measured from architects' drawings and then encoded to a computer system.

The more fundamental research task was to develop an interface between architect and computer, and data structures within the computer, such that the architect's intentions could be adequately represented within the computer. The quantification of construction information associated with design descriptions and the interface with quantity surveying practices were not expected to present major new problems. However, development of computerised quantification techniques, as a peripheral adjunct to the research, brought researchers into contact with quantity surveyors. The subsequent efforts to relate the concepts underlying computer systems to real world practices as perceived by quantity surveyors provides the context for our further discussion.

Quantity surveying practices

Quantity surveyors translate design information presented by architects, in the form of drawings, into quantified descriptions of the construction necessary to execute designs. They represent this information in written schedules and bills of quantities. The purpose of quantification may be summarised as follows:

(a) to provide a basis for estimating probable costs, to advise the architect during the development of a proposed design;
(b) to provide a common basis for assessing price estimates from different building contractors at the time when tenders are submitted for a building contract;
(c) to provide a basis for controlling costs and regulating payments during the execution of a building contract.

Quantity surveyors commonly perform an interpretive role, adding missing information to an architect's drawings and compiling information on building operations. They serve as an interface between architect and builder. In this role, quantity surveyors have to lean heavily on their knowledge of the building industry. In addition, their practices are regulated through use of the Standard Method of Measurement, or SMM.

SMM seeks to provide a procedure for identifying and quantifying construction information, with extensive references to instances of construction. It is relevant to note, however, that SMM contains very little method that is independent of instances. Further, it is significant that where any conflict occurs between SMM and a quantity surveyor's own experience, the quantity surveyor will prefer to rely on his or her own experience.

SMM implies a particular building model: an aggregation of lots of separately defined parts of construction constituting a whole building. An underlying assumption is that by finer decomposition into separate parts, we get more accurate cost associated with each part and, through summation of parts, we get a more accurate total cost. Architects may use other building models. For example, an architect might think of a building as an aesthetic expression of some human aspiration which the building is to house, the whole and the parts then being viewed in entirely different terms from those anticipated by SMM. The ensuing negotiations between architect and quantity surveyor implicitly seek to reconcile their respective building models. Slowly, their separate perceptions of buildings may change. Parts of SMM may be found to be impractical and some of the architect's ambitions may be found to be unrealistic; building models become adjusted.

Understanding models
Such reconciliation between different models also comes into play when any practitioners are presented with computer applications. The model used by the SSHA system employed a concept of parts which mapped onto the three kinds of entities: junctions, components, and spaces. To achieve this mapping, entity types had to

Integrated Design Systems

be defined in a manner that would clearly differentiate one from another and, consequently, parts of buildings had to be understood in terms of precise, albeit abstract, boundaries.

The quantity surveyors were already employing a concept of modules of quantities associated with construction details. These modules were read off architects' detail drawings. Quantity surveyors could recognise parts depicted in a drawing and quantify them. They could also recognise features of other parts only partially depicted in the drawing and quantify the operations necessary to produce the features. Figure 4.4 shows an example of the latter, in which a drawing depicts a junction between a window and a wall. The drawing depicts junction parts plus partial depictions of the wall and window parts. The wall part shows the closure of a brick cavity to form the reveal for the window. Closure of the cavity would be quantified as an operation associated with the junction, without affecting the material quantity measured separately for the wall. Thus, we get a junction module that includes some part of an adjoining wall, co-existing with a unit measure of wall that ignores the same part.

Architects' detail drawings generally did not reveal demarcations beween junction parts and adjoining components, and the partial depiction of each component usually faded into some undefined length of component. Quantity surveyors' interpretations of these drawings into junction modules depended on their recognition of material parts and operational features which, in turn, depended on their knowledge of building construction.

A computer system could not possess equivalent powers of recognition. The problem then was that neither architects' drawings nor quantity surveyors' modules were sufficiently explicit to be represented directly as entity types within the SSHA system. To supply the system with a library of construction details, quantity surveyors had to become familiar with the model's notional boundary to junction entities and learn how to be explicit in presenting separate quantities associated with junctions. Architects and quantity surveyors then had to develop a set of precisely bounded junction details for all anticipated combinations of components in any of the four possible geometric configurations; end-on, right angled, tee, or cross-over. A sample of such details is shown in Figure 4.6. Here we have an illustration of the kind of adjustments that practitioners have to make to their own understanding of their practices, in order to represent some part of their practices as a model in a computer system. To do so, the practitioner has to become familiar with the model.

Despite similarity between the model's junction entities and the quantity surveyors' previously established modules, it cost the quantity surveyors and the architects a lot of effort to rework their practices in order to create the system's library of construction details. At the time, these practitioners did not yet have any experience of the system which they regarded, somewhat sceptically, as being rather theoretical. After it became operational, the prior existence of quantity modules remained important to the quantity surveyors' eventual acceptance of this system.

Users' preconditions
The quantity surveyors had no previous experience of CAD applications involving computer graphics. Not being in a position to recognise or evaluate technical aspects of formal procedures programmed in a computer, these quantity surveyors set out three primary conditions to be met before they would accept the system. These conditions were:

(*a*) *totality* — the programs must cover complete jobs: buildings, or site layouts;

(*b*) *accuracy* — results must match the accuracy achieved by pre-existing quantity surveying methods;

(*c*) *procedures* — CAD procedures must be visible and retraceable to enable quantity surveyors to check results.

On the face of it, these appeared to be quite plausible conditions. But during subsequent discussions, consideration of existing quantity surveying practices in relation to these conditions threatened to undo the researchers' attempts to develop a corresponding computer model.

(a) Totality:
The concept of totality, for purposes of defining the boundary to a job, proved elusive. Recognition of the completeness of a job depends on a quantity surveyor's perception of the job, a building, or a site layout. The nature of this perception was hidden within individual quantity surveyors. It was not possible to discover a consistent procedure that would ensure a complete description of all the parts which might constitute a whole job, which would cover all that a quantity surveyor might see — parts might be left out and other parts notionally added. As an example, in the case of steel reinforcement to 'no-fines' concrete construction, in certain situations (around windows along eaves walls) quantity surveyors would be concerned to establish the form and extent of reinforcement in great detail, whereas in other situations (in gable walls) the

existence of reinforcement would be ignored. Again, in the case of services, electrical wiring was ignored whereas plumbing pipe runs tended to be measured in great detail. There were usually sound practical reasons for such inconsistencies.

Yet, in demanding that CAD programs must cover complete jobs, quantity surveyors quite reasonably sought to avoid being landed with the task of identifying demarcations between CAD and manually measured quantities, and the task of amalgamating these quantities. Such tasks can lead to ambiguous situations and awkward questions about responsibility. They would also require the quantity surveyors to possess full knowledge of operations within a computer.

(b) Accuracy:
The demand for accuracy in CAD programs prompted questions about the accuracy of existing manual practices. Accuracy of manual practices was variously quoted as between 2 per cent and 4 per cent, without specifying how the percentage was measured. Quantities as a percentage of building built cannot be tested; comparison of quantities taken from drawings and from actual buildings is likely to involve self-cancelling errors, and differences might be attributable to differences between perceptions of quantity surveyors. Comparison between a cost estimate and contract price would, in practice, need to be qualified by uncontrolled events that take place between the time of estimate and time of tender (typically, three or more years), and such comparison would be obscured by the variations in tendered prices. There is no absolute way of testing the accuracy of tendered prices.

As already noted, there is a general assumption underlying quantity surveying practices, which is that improved accuracy is obtained from increasingly detailed measurements. Thus, breaking down a total job into minute and separate component parts is believed to ensure more accurate global quantities and, therefore, more accurate cost estimates.

It was believed that accuracy of measurement could be tested by comparing CAD produced quantities with quantity surveyors' quantities for the same job. Any differences were presumed to indicate inaccuracy or error on the part of computer programs. In one example, when testing earthwork quantities associated with road alignment, produced by the site layout system, several days were spent by researchers searching for an error in the program before it was discovered that the quantity surveyor had made an error in manual measurement.

(c) Procedures:
Existing procedures for quantifying and costing design proposals, for purposes of comparing alternatives before proceeding with detail design, could not be identified as overtly describable procedures that were executed consistently by different quantity surveyors. In the case of site layout, it appeared that any given quantity surveyor would look at a given plan and spontaneously come to a global assessment: the site is expensive, or cheap, or somewhere in between. He or she would then measure selected elements of the plan and moderate the associated cost rates according to the global assessment. Summing the parts, the result would corroborate the initial assessment. The result might be accurate, depending on his or her ability to call on personal experience of the building industry, and this procedure can be regarded as very efficient. But the procedure remains idiosyncratic to quantity surveyors.

For the computer system, it was necessary to develop programs to quantify all the elements that constitute a site layout, including all the attributes that contribute to the complexity of a layout. To arrive at a cost estimate, it was necessary to associate appropriate costs with these quantities, taken from a cost library maintained by quantity surveyors. These programs had to process complete quantities to arrive at a corresponding global estimate. This process was similar to that used by quantity surveyors for preparing full quantities, and advantage could be gained only by operating the process very fast within a computer. Acceptance of results by quantity surveyors was then conditioned by the questions of totality and accuracy that have already been outlined.

Persuasion and acceptance
During discussions with quantity surveyors it became apparent that method, in the sense of procedure that can be defined independently of particular instances, plays a small part in determining the outcomes of quantity surveying practices. These practices are more dependent on knowledge of a vast body of special cases, continuously adjusted by new experience.

The observation that quantity surveying practices depend on knowledge which remains within quantity surveyors, coupled with the researchers' inability to formulate an explicit model for any whole view of quantity surveying practices, poses an odd question. How can different quantity surveyors believe each other? The visible products of their work imply a precision which invites this question. The answer, quite simply, is that confidence among

Integrated Design Systems

quantity surveyors is built on their common background and experience, having been trained at the same schools. They understand each other implicitly, so they agree explicitly to acknowledge the validity of each other's products.

Being involved in developing computer programs for quantifying design information, the researchers then found that they had a major problem. Their perception of quantity surveying practices, motivated by their commitment to computing, could not coincide with quantity surveyors' experience or their own practices. The quantity surveyors' model, whether perceived by themselves consciously or implicitly, could not be the same as the separate model which the researchers were actively creating through their different perception.

The position was still worse. Quantity surveyors saw no need to formulate an explicit and all-embracing model of their activity. They do not have to externalise the totality of their practices. Instead, they can call on those abilities that remain within themselves as members of a community of like-minded persons. Therefore, the researchers had no basis for demonstrating to the quantity surveyors that their explicit and externalised model, embodied in computer programs, was compatible with the quantity surveyors' own practices. The researchers could not expect the quantity surveyors to acknowledge compatibility between the researchers' model and the quantity surveyors' own understanding of their practices. Indeed, the very concept of models was elusive to them. The researchers could not reason their case.

It was necessary, therefore, to find alternative means for persuading quantity surveyors to accept CAD programs in their practices. It became necessary to enter into a long and informal series of discussions, to build up understanding between people, somewhat independently of the technical considerations associated with the programs. This process continued over years, and required the help of other developments unconnected with CAD. The incentive to continue this process did not come from the quantity surveyors but was imposed by pressures originating from other parts of the user organisation, and from outside the organisation. Changes in quantity surveying practices arose from increased demand for fast response to calls for help from architects, in a general situation of increasing workloads. This increase was not expressed in terms of more buildings, but was the result of more questions being raised from other people about the design and cost of buildings. These pressures prompted further rationalisation of existing practices.

As an example, we can refer to the problem of measuring gable/mutual walls between houses, in site layouts. At the time when the SSHA system was being implemented, ground forms of housing sites were becoming more irregular, as the more level sites for urban development were becoming scarcer. In consequence, site plans included more vertical steps and horizontal staggers at the junctions between houses, in order to mould terraces of houses to the ground. Quantity surveyors could rationalise the superstructure information contained between party walls, which remained relatively stable irrespective of how houses were arranged on site. However, they were faced with a need to look at the junctions between each pair of houses to identify and quantify the unique information occurring at each gable/mutual wall. The labour involved in this was considered to be disproportionate to the work of quantifying the rest of each house description.

To reduce the workload on measuring gable/mutual walls which would occur anew for every pair of houses on every new site, the quantity surveyors proposed to standardise permissible dimensions for steps and staggers between houses. Architects resisted this proposal, seeing it as threatening the last domain of freedom that allowed them to respond sensitively to site conditions and other design considerations. The quantity surveyors also found that their own proposal presented themselves with a worrying problem. Their intention was that standard steps and staggers should result in standard gable/mutual walls which could be quantified and stored in a library for use on different site layouts. However, even with a restricted set of dimensions for steps and staggers, they found that the possible permutations between different house types resulted in thousands of 'standard' gable mutual walls which they would have to quantify and store. This was a daunting workload. In this situation the quantity surveyors became more receptive to computer programs which could look at all instances of houses and describe and quantify every gable/mutual wall between every pair of houses on any site, without dimensional constraints on the steps or staggers. This became one of those rare moments when the architects and the quantity surveyors could both sense that they would obtain real benefit from a CAD application.

Eventually the quantity surveyors, without necessarily having explicit technical knowledge of CAD, came to believe in the practicality of using the quantification programs which formed part of the larger SSHA system. Even after having succeeded in gaining acceptance by quantity surveyors, there is no methodical way of evaluating this achievement. In the absence of any other criteria, we

Integrated Design Systems

can observe the evidence of benefit experienced by the user organisation and acknowledge its commitment to CAD.

Software Technology

We have now discussed some of the questions that arise from any conjunction of users' work-practices with a computer system, focusing on just one activity within a design organisation. Having focused on the work of quantity surveyors, we should expect that the questions which have been posed have even greater significance for architects and for designers in general. These questions become critical in the case of integrated design systems as defined at the start of this chapter. The expectation that a computer should know about a designer's world, and that it should be able to use this knowledge to respond to whatever it is told by the designer, means that the system has to be modelled very closely to the designer's own perception of design. The issue for software technology then is whether the technology can be made to reflect different designers' varied and changing perceptions of design. The following paragraphs will illustrate connections between software technology and the SSHA experience.

Prescriptive technology

The SSHA system used what is now considered to be old or established technology (FORTRAN). This technology requires programmers to have detailed knowledge of available computer operations for storing and accessing data and for executing arithmetic functions on data. Control of these operations is gained through the use of the programming language. The programmer tells the computer where data are stored by reference to locations (addresses) in machine memory, and tells the computer which operations to apply to data. By assembling many such instructions, the computer can be programmed to exhibit complex behaviour which users might recognise as meaningful to their own view of an application domain.

This general strategy is typified as procedural access to data by reference to location, in contrast to newer techniques which support declarative access by reference to information content (as when naming things).[4] Programmers get what they want by telling a computer how to execute procedures rather than by telling it what they want. As a general condition, procedures have to be defined as complete series of sequential operations which determine singular outcomes. When modifications have to be made to already programmed procedures, they have to be inserted in such a manner as to ensure that this condition remains satisfied. Here we come to the problem presented by this strategy: the knowledge used to write a program in this way is highly complex; it is not easily retained by the author; and it is not easily recoverable by another programmer reading the program code. More generally, the knowledge required to write a program is highly specialised and far removed from the knowledge of most computer users.

A designer's access to computer operations depends on a specialist serving as interpreter between designer and computer. The specialist programmer looks at a user's world and translates what he or she sees into a system that he or she knows can be implemented as a set of machine operations. The designer's access to the full potential of the computer is limited by the individual programmer's knowledge of computers. More problematically, the programmer's task in developing a program is dependent on the programmer making predictions about the designer's use of the completed program. These predictions can be based only on the programmer's view of the design domain. A program will take time to develop. The designer's work-practices will evolve during that time. The introduction of a computer into the work-practices of the designer will cause those practices to change. For the finished program to function at all, irrespective of whether the designer also finds the program useful, the designer has then to conform to the predictions which the programmer has built into the computer program. The designer has to supply data and invoke procedures precisely in the manner prescribed by the programmer. This need to comply with a programmer's predictions characterises older software technology as prescriptive technology.

Most currently established CAD systems which perform tasks specific to users' application domains rest on prescriptive software technology. The effects of prescriptiveness may be masked by a good user interface which might ensure that the user is aware of the expectations built into the system. A good interface should also guide the user through available options offered by the system at any stage during a program execution. But, however good the user interface is, the designer will not be able to step outside the bounds of the computer application as conceived and implemented by the programmer.

System implementation

The ideas underlying the SSHA system were ambitious and the system took several years to develop. This work employed the efforts of many programmers and a succession of programmers. The system became very large; by the time it was put into practice, there was no longer any one person who knew how the whole system worked.

As already noted, a computer system is the product of a programmer's interpretation of his or her perception of a user's domain in terms of procedures that he or she knows can be implemented in a computer. By multiplying the number of programmers on a single project, there is inevitably a risk of confusion arising from their various knowledge of computers, and from their different perceptions of the user domain. As a system grows, and as programmers are replaced, it becomes increasingly difficult to undo predictions that are already built into programs. Changes to any part of program code increasingly produce unforeseen ripple effects throughout the rest of the program. Thus, early decisions tend to become entrenched and unalterable in a system. When a system takes several years to develop, eventual users of the system have to live with those early decisions.

The SSHA system was developed on an organisational strategy for data that was supportable by available technology, which took the form of hierarchical records and pointers structures. Essentially, records consist of data plus addresses which serve as pointers to locations of other separate data. Data structures consisting of linked records can then represent logical structures which correspond to assemblies of parts describing design objects. Pointers usually represent only one kind of relationship between parts: a 'part of' relationship. This means simply that one item of data forms part of another item of data within some hierarchical arrangement of records. The SSHA system used a variation of the regular tree hierarchy, in the form of an interlinked ring structure. Rings were used to represent component boundaries and room space boundaries as shown in Figure 4.3. These rings implied a 'next to' relationship, and whole rings were interlinked through the usual 'part of' relationship. Composite structures of this kind, subject to detail variations and new names, are still in use today.

The issue here is that any such structuring of data implies a particular view of a user world as represented in a system. The structure defines access paths to data and, therefore, conditions what users can do with a system. Any change in the user's world, which was not anticipated in the design of the structure, leads to demands for the structure to be changed. New access paths require changes to the structure, and such changes require the assistance of a programmer. When the user requires such help, the programmer may not be available. In the example of the SSHA system, programmers with the necessary knowledge of this system no longer existed.

To illustrate this point, we can look at the simple case of the SSHA system's representation for components. In general, components were considered as slabs of layered materials, with edge conditions described separately as junctions. Construction could therefore be described as materials assigned to identified layers of a component. Figures 4.9 and 4.10 show how this information was represented in a hierarchical data structure. This structure had a master list of records identifying junction, component, and room types. The figures illustrate further information associated with a wall component, contained in further records identifying the primary construction and its candidate range of materials, and external and internal cladding with their respective ranges of materials. This was regarded as a plausible representation for components, and became deeply embedded in the system. However, this representation had limitations.

This structure presented a view of components in which external and internal cladding materials were independent of each other and were separately dependent on the choice of materials for the primary construction. There was no provision for surface finishes that might be dependent on choice of materials for cladding. More seriously, there was no provision for identifying separate layers of primary materials that might be variously dependent on each other. Brick plus cavity plus block construction had to be represented as a single composite materials specification.

This is a simple example. Consider the more elaborate relationships shown in Figures 4.3, 4.5, and 4.7, where the data structure had to represent arrangements of components linked through junctions to other components and through surfaces to room spaces. These relationships were subject to similar limitations, so deeply embedded in the system that they remained unaltered for more than ten years.

The problem illustrated here is that established software technology requires design knowledge to be represented by data structures embedded in a system, in a manner which cannot readily be undone. Following decisions made by programmers early in the development of a computer system, eventual users are far too dependent on the programmers having been right. In the field of design, this dependence on correct predictions is fatal. Necessary *ad hoc* modifications eventually make a system unmanageable and it dies. This is a fundamental criticism that can be overcome only by radical changes in the concepts employed by system designers and users of computer technology.

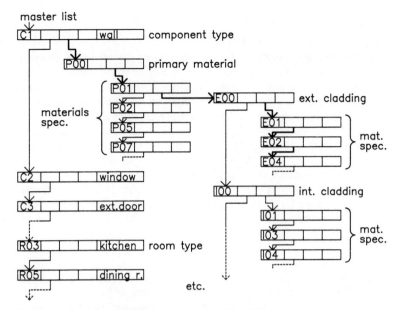

Figure 4.9 *Records and pointers data structures*. The SSHA system employed a hierarchical decomposition of components into primary materials and cladding materials, differentiating between internal and external cladding in the case of external components. This model was represented by data records with pointers to the locations of other records. In implementation, the model became deeply embedded in program code so that this three-layer view of components became entrenched in the system.

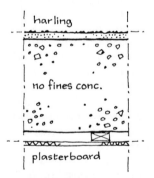

Figure 4.10 *Wall Components*. The hierarchical decomposition was used to represent components, such as this example of a no-fines concrete wall with plasterboard dry lining internally and harling for its external cladding. The representation maintained connections between these three layers of materials so that permissible alternatives could be identified if any part of the specification was changed. Such changes might be initiated from the room space bounded by the component, and could propagate changes through junctions to other components.

Lessons from Experience

We have now looked at two examples of integrated design systems. We have looked in some detail at one of these systems, the user organisation for which it was developed, the experience of users within that organisation, and the implications of the technology on which the system was implemented. What does all this tell us?

First, we should recall that the SSHA system was conceived twenty years ago. Despite its age, the experience it has provided and the issues which that experience has exposed remain highly relevant to present day ambitions for CAD. The SSHA was an adventurous pioneer in applying CAD, in making CAD productive, and in making CAD financially successful. Now our purpose is to draw on this experience, to inform ourselves on the possibilities and limitations of future more generalised CAD systems.

User organisations

The Scottish Special Housing Association presented a highly sophisticated organisation that exhibited regularities to a far greater extent than is usual in design offices. It sponsored the development of the SSHA system and, consequently, these irregularities provided a tangible basis for the specification of a practical system. The existence of these regularities increased the probability that the system would prove useful in practice. However, the advantages that the Association was able to offer are not likely to be repeated.

The very substantial benefits experienced by the SSHA were largely attributable to the CAD system's ability to interpret drawings in terms of other non-graphical information about buildings. This strengthened the Association's position in its dealings with client and construction markets. However, the interpretive ability of the system was dependent on the existence of regularities in the Association's normal work-practices.

Even in the example of the SSHA, the regularities of its design practices proved fragile. Over time, given national changes in demand for housing, coupled with changes in political commitment to semi-governmental agencies with executive responsibility, the role of the SSHA changed. The two main effects were a reduction in the scale of the Association's activities and a shift in emphasis from new housing to the modernisation of old stock. The SSHA system could not be modified to meet changing demands from the Association's own changing design practices. The system became a brittle core within an organisation that needed to be flexible and, eventually, after more than ten years of use, the system had to be discarded.

Integrated Design Systems

New user organisations are going to want CAD systems which are far less dependent on in-built models corresponding to particular design practices. But they will want the ability to relate graphical and non-graphical expressions of their own design knowledge, in a manner that can reflect any changes to their own design practices. To meet this criterion, future user organisations will want systems which offer users far greater control of computer operations for describing and performing tasks on descriptions of design objects. This need for control by users ought to take precedence over demands for sophisticated and domain-specific functionality pre-programmed in a system. This need for control is essential for user organisations engaged in evolving design practices.

Users' practices
People doing practical things cannot be bound by some wholly explicit and systematic representation of their tasks in computers. They also need to call on their own intuitive abilities as people. This point has been illustrated by the example of quantity surveyors' practices. That this should be true of practices aimed primarily at the task of quantification means that this point has to be accepted as highly significant for all other design tasks.

Computer systems must be able to accommodate demands and responses prompted by users' own intuitions. This requirement poses fundamental questions about the formalisation of tasks and boundaries between tasks, and about the extent to which human abilities should be represented in machines. These questions cannot be resolved solely by technological advances.

Future users will need to claim greater responsibility for formulating tasks in computers, and reformulating those tasks as demands from outside a system change. To do so, they will have to become familiar with abstract modelling techniques and with the use of logical relationships in models. It is only then that they will be able to use computers to support their own design tasks and communicate results to other people. We will return to this issue in Chapter 8, where I consider the broader implications of computer literacy.

Use of technology
Currently established prescriptive technology cannot meet the practical needs of designers. Instead, we need to recognise an affinity between the activities of developing programs and designing things. In both cases, acceptability of results is dependent on the perceptions of people who receive them — the users. There are no single, demonstrably correct solutions to programs. A particular

difficulty presented by programs is that people, as users, cannot readily visualise them. This general position was illustrated by the researchers' efforts to persuade quantity surveyors to accept the SSHA system.

The prescriptive technology used by the SSHA system led directly to the system's demise. It was the rigidity of its complex data structures, the need for close correspondence between these structures and users' own demands on applications, and the inability to change the structures as demands changed which eventually made this system unacceptable. More recent advances in software technology have sought to distance computer operations on data from the users' view of applications by interposing some generalised logical representation scheme. Operations that support the representation scheme can then be invoked implicitly for various applications and reinvoked as users' needs change. Such use of computers requires knowledge of logical relationships, with little regard to deeper operations within a computer. In Chapter 7 I will develop an example of this approach to CAD.

SUMMARY

After outlining two early examples of large integrated design systems, OXSYS and the SSHA system, this chapter has focused on users' experience of the latter system. The extensive scope of these systems, in terms of their coverage of users' design practices, makes that experience relevant to current ambitions for CAD. The noteworthy achievement of these systems is that they were employed productively within design offices over a period of more than ten years. Despite the success of these two systems, they did not mark the start of a general acceptance of integrated design systems in design offices. The discussion of experience explains why.

The sponsor and user of the SSHA system has been described in terms of the regularity of its design practices, its expectations of the CAD system, and the benefits it obtained from CAD. We then looked more closely at the experience of quantity surveyors within the user organisation, as users of the SSHA system. Despite their emphasis on quantification tasks, it was found that even they conduct practices which do not readily decompose into tasks which can be detached from people and represented in computers. We should not expect computer assistance to designers to be any easier. Acceptance of CAD required persuasive efforts of researchers, as between people, and was induced by circumstances within and outside the user organisation. Significantly, the successful exploita-

Integrated Design Systems

tion of the system was dependent on the prior existence of highly rationalised practices within the SSHA.

This experience was related to general considerations of software technology, at the task of telling computers what users want them to do. Dependence on specialist programmers who translate their perceptions of user domains into computer operations, and on programmers who encapsulate predictions of future users' actions in program code, typifies currently established software technology as prescriptive technology. We need to devise systems that allow users far more control over computer operations that are far more responsive to varied and changing demands in the world of designers. This ambition might seem unreasonable to some researchers, but it gains importance from being rooted in practical experience, and designers need answers.

Notes to Chapter 4
1. OXSYS is described by Hoskins (1977); and the SSHA system is described by Bijl et al. (1970), and Bijl (1978).
2. Since this chapter was written, the SSHA has been amalgamated with the Housing Corporation in Scotland (HCIS) in a new body called Scottish Homes, on a path towards privatisation.
3. This housing site layout system is described by Bijl and Shawcross (1975) and Bijl (1979a), and the work on polygon set operations is described by Holmes (1978).
4. Early development of declarative database access techniques is described by Date (1975), and a more recent review is provided by Frost (1986).

CHAPTER FIVE

FUNCTION-ORIENTATED SYSTEMS

If I ask you to do something, you might ask me why, in order that you might get to know more about what I want: we need to consider what happens when computers are asked to do useful things, and whether they can get to know what we want.

Now we come to function-orientated or task-specific systems. The initial assumption that prompted this approach to CAD was that computers should do things, and that the actions of computers should be useful to the tasks people do. This assumption became an ambition, and the technology available in the 1960s and 1970s set boundaries to this ambition. Tasks had to receive complete overt descriptions, and they had to be viewed as being independent of each other, as discrete tasks. Tasks also had to be repetitive, to be recognised and performed in the same way by different people over long periods of time, to ensure benefit from investment in program development. With these qualifications, task-specific systems were viewed as an attainable goal for CAD and they led to a plethora of small-scale and large-scale computer programs.

From our discussion of design and integrated design systems in Chapters 3 and 4, it should be no surprise that function-orientated systems have failed to find widespread application in design practice. This is particularly true in the field of architecture. The problems facing these systems arise from the discreteness of tasks and the need for common recognition of tasks among different practitioners, coupled with the difficulty of representing changing human perceptions within established computer programs. These problems rest on philosophical stances of the kind aired in Chapters 1 and 2, pointing to the relevance of philosophical considerations for day-to-day practical activities. In this chapter I will discuss some

Function-Orientated Systems

of the issues presented by experience of function-orientated systems.

Task-specific Systems

A function-orientated system is a system in which an anticipation of some specific task in a user's world forms the primary motivation for developing the computer system. An example might be the calculations required for environmental appraisals to evaluate the performance of proposed designs for buildings.[1] The anticipation of task is paramount in providing a specification for required system operations, and the organisation of data is regarded as subservient to the task. Typically, these tasks each require separate input and provide output which users have to translate into their own perceptions of design.

The general model of design implied by this function-orientated approach may be summarised as:

(*a*) designing is a process which consists of tasks that people apply to things;

(*b*) any whole design can be analysed into discrete parts which take their definitions from bounded tasks;

(*c*) decomposed parts can be categorised as quantitative, amenable to computer processes; and qualitative, subject to human judgement and decision;

(*d*) design products are some kind of summation of the results of tasks applied to parts.

The philosophy underlying this approach to systems is illustrated in Figure 5.1.[2] Here, designs are regarded as solutions to problems and designing is seen as a kind of problem-solving process involving analysis, synthesis, and appraisal. Analysis refers to the decomposition of problems into parts which we know overtly how to treat and which are computable; and other parts which we treat intuitively. Synthesis refers to a fusion of parts into a whole which is more than the product of aggregation or composition, which remains partially unexplained. Appraisal refers to the evaluations of parts which are defined by measurable goals and which are computable; and other parts or aspects which are subject to human judgement. Thus, computers are seen as having a role in analysis and appraisal, and humans are responsible for synthesis, taking due account of outcomes from computers. The figure shows this process as being cyclic, with results from computer appraisals prompting modifications to the designer's synthesis. If we then also view designs as expressions of intentions for designed artefacts in the form of words (in design briefs and building specifications or bills of

Computer Discipline and Design Practice

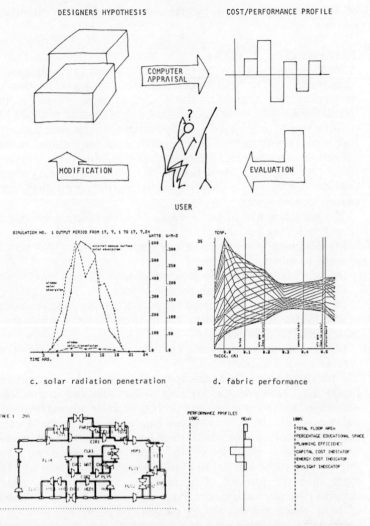

Figure 5.1 *An approach to function-orientated systems*. A cyclical approach to design is illustrated (above), in which designs are hypothesised by people and presented to computers for appraisal. Results of appraisals are presented as cost/performances evaluations, back to people (below). These evaluations then prompt modifications to design hypotheses. Examples of computer appraisals in the form of graphs (middle) illustrate aspects of energy performance, produced by a sophisticated thermal energy simulation program.

Figure by permission of ABACUS, University of Strathclyde.

Function-Orientated Systems

quantities) and in drawings, then it can be argued that the analysis/synthesis/appraisal cycle and the role of modification applies to all stages of designing, even initial designs. This approach to design was strongly advocated throughout the 1970s and survives into the 1980s.

The function-orientated approach requires discreteness of decomposed parts of design in order that the parts can be entered into a computer and be subjected to quantitative functions. The major simplification over the integrated design approach is that humans remain responsible for the integrity and consistency of a whole design, exercising judgement and decision on the results of calculations. The scope of computer programs is then limited to particular anticipations of quantitative tasks, each handling only those data and operations appropriate to a discrete task.

Problems arise from the boundedness of discrete parts and the consequent notions of classification and typing within a user's domain, which imply particular perceptions of design. Boundaries between one task and another vary for different designers and, more seriously, boundaries between quantitative and qualitative tasks do not remain stable for different instances of design. Previous discussion of the practices of quantity surveyors, in Chapter 4, adds substance to this point. In that experience, it took a long time both for computer programmers and end users to recognise that quantification serves as a means of expression which encapsulates qualitative considerations.[3] Quantification is employed in numerical formulations which are used to represent particular human perceptions. When perceptions are changed, existing formulations become contentious. Design embraces changing perceptions.

EXAMPLE OF GROUND MODELLING

To consider the practical implications of the function-orientated approach to CAD, we can look at an example of ground modelling.[4] The objective of ground modelling is to produce a description of any ground surface. The description can then be used to evaluate effects of proposed changes to the surface, like levelling off parts for roads and buildings. Applications typically include the calculation of earthworks quantities.

The strategy employed for this ground modelling program was to start with random spot height data as digitised input. These data were then used to interpolate height data for the nodes of an automatically generated grid, using inverse square weighted averages. The surface model was then produced by applying a quadratic least squares surface fitting algorithm to the height values at the grid nodes, moderated by the actual input values. The model was made visible by generating ground height contours, employing a curve smoothing algorithm and including a routine for clipping contours to site boundaries. The techniques employed were not radically new. The aim of this work was to explore the effects of new database strategies employing AT&T's Unix environment on program implementations. Compared with previous FORTRAN experience, improvements in ease and speed of implementation were impressive. However, response from users to the program proved interesting.

Early tests of the initial program by architects produced results which were computationally correct but alarmingly different from users' expectations. One reason was the differences in knowledge which users and the programmers associated with contours. Architects are used to regarding vacillating contours as an indication of uneven ground, whereas the program tended to show such contours meandering across relatively flat ground. In addition, the surface fitting algorithm tended to smooth out sharp changes in ground surface.

A major difficulty was that users came to the program with a particular image of a ground surface in mind, and they would enter spot heights from hand-drawn contours expecting to see the same contours reappear on the display screen. They found difficulty in appreciating that any given (sparse and uneven) set of spot heights can generate different 'correct' models and contours, depending on the fine tuning of procedures employed by the program. These issues were not made any easier when control exercises resulted in hand-drawn contours which were found to be in conflict with spot height survey data. Student architects who took part in these exercises were asked why their contours sometimes passed on the wrong side of spot height values. They answered that they knew how the ground went — they had been there and seen it — the measured values must have been wrong. The users 'knew' what the surface was and its explicit representation as contours had to conform.

Further development of the program entailed extensive negotiations between users and programmers to identify and alter procedures in the program, to achieve acceptable ground models. The following list is given only as an indication of the kind of issues that became subject to negotiation:

(*a*) the procedure for establishing the grid, varying its size;
(*b*) the number of spot heights used to establish grid node heights;

Function-Orientated Systems

(c) the effect of distance between spot heights and grid nodes;

(d) the weighting given to spot heights over grid nodes when executing the surface fitting procedure;

(e) extent of ground area influenced by spot heights.

Negotiations also had to clarify the definitions and effects of:

(f) sparse data;

(g) uneven distribution of data.

The final version of the program included some significant changes over the initial version:

(a) the size of the grid was reduced, for a given number of spot heights;

(b) in the initial program, grid nodes were simply assigned their nearest spot height values as node start values, subsequently modified by surface fitting iterations — later versions of the program introduced an inverse square/distance weighting algorithm operating on numbers of spot heights — the final version determined node start values by defining a 'catchment' area for each grid node and then applying the inverse squares algorithm to all the spot heights within the catchment area;

(c) weighting of spot height values was reduced considerably for purposes of carrying out surface fitting, and the area of influence of spot heights was also reduced — the number of surface fitting iterations necessary to produce a satisfactory ground model varied, depending on the form of the ground surface and the density and distribution of spot height data, but generally two iterations were found to be sufficient.

These considerations may appear a bit technical and it is not important to understand or verify them. The point is that they all involved negotiations which rested on judgements by programmers and users. Examples of the effects of changes implemented during the course of negotiations are given in Figures 5.2, 5.3, and 5.4.

Acceptability of the task

Ground models are used to represent actual pieces of ground which exhibit naturally formed surfaces, unconstrained by any formal geometric concepts. From a computational point of view, the surfaces are arbitrary. A central problem in developing procedures for modelling ground is that the visible presentation of the computer's model, such as a display of ground height contours, must be recognisable as matching the user's own view of the ground. The problem is twofold:

(a) any procedure operating in an internally consistent manner

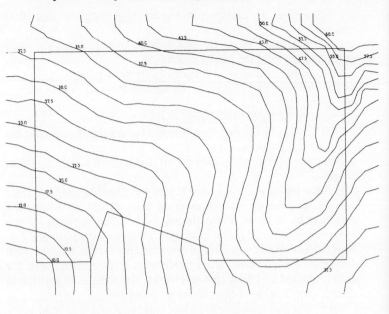

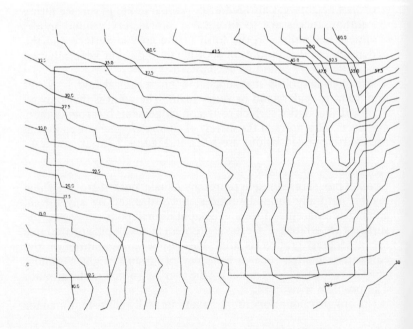

Function-Orientated Systems 123

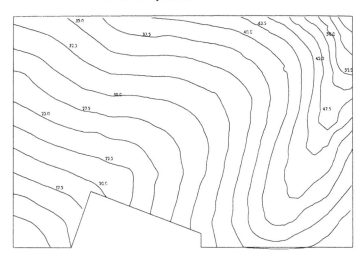

Figure 5.2 *Development of a ground model*. Examples of interactions between programmers and users during later stages of program development. The computer generated contours (bottom left) show the result of adequate spot height data and surface fitting to grid node height values, using a procedure for finding initial node values from the weighted average of six nearest spot heights to each node. Users were unhappy with the spikiness of contours. The contours (top left) are the result of a different procedure, using inverse square/distance averaging of all spot heights within a 'catchment' area around each grid node, producing 'better' contours even before surface fitting. Users were happier, though there was still some spiking. The contours (right) show the effect of two surface fitting iterations to initial grid node values found by the 'catchment' procedure, and the contours have been smoothed and clipped to the site boundary. Users were satisfied. However, there remained no way of knowing that these were the correct set of controus among all other possible contours that could reasonably be produced from the same input data.

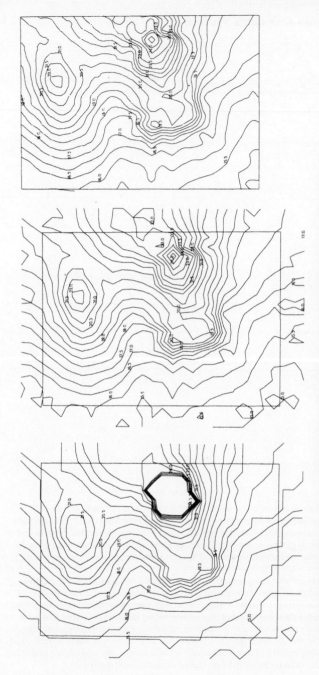

Figure 5.3 *An example of more irregular ground.* This series of contour maps shows a progression of contours before surface fitting (left), after surface fitting (middle), and after smoothing and clipping contours to the site boundary. Note that the bunching of contours around the peak (left) has been resolved by the surface fitting procedure.

Function-Orientated Systems

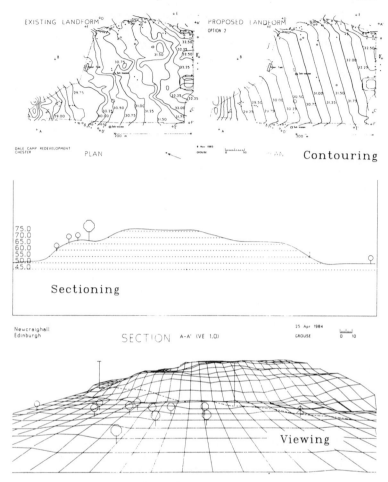

Figure 5.4 *The modelling program in use*. The previous figures show the program when it was first being developed, over a period of a few months. This figure illustrates the program after further years of development and use in practice by landscape architects. Continuous interaction between programmers and users was required to refine the model, to maintain confidence in the quality of contours, and to extend applications to include sectioning and three-dimensional surface projections.

will have practical limitations on the arbitrariness of the surfaces it can represent — it can only impose a surface it 'knows' on the available information supplied from the actual piece of ground; (b) it is practically impossible to make the user's view of a piece of ground wholly explicit — any explicit presentation, such as a table of spot height values, will be supplemented by the user's own direct experience of the ground, which tends to remain inside the user's mind.

This problem cannot be resolved simply by writing bigger and better programs, or by demanding that the user makes all knowledge explicit. Even if bigger programs can be afforded which might, for example, implement a more sophisticated surface fitting algorithm, users would still be unable to provide all the necessary data. The distinction between explicit and intuitive knowledge describing a piece of ground may not be consciously appreciated by users. It is usual for the user to discover that he or she knows something more only when faced with a surface representation which he or she mistrusts, and only then will be prompted to reveal that further information to explain why the representation is wrong.

We can make a further, uncomfortable observation. The accuracy of a ground model, in terms of its accurate representation of an actual piece of ground, cannot be measured. There are no objective references for such measurement. Instead, accuracy has to be considered in terms of compatibility between the separate models created in a computer and in the mind of a user. A computer model can be correct, in terms of procedures correctly operating on given data, but acceptance of such accuracy remains a matter for judgement by users. The different intentions of designers and others who use ground models, for different pieces of ground, result in changing criteria for acceptability. In practice, the development and use of a computer program to generate ground models is subject to repeated negotiations between user and programmer, leading to frequent program modification.

Intractability of tasks

Ground modelling has been discussed as a computerised task that is intended to aid designers when they design things. It has been described as a discrete task, operating on spot height data to produce ground surfaces. As one example of task-specific systems, it differs from other tasks in its particular functionality. Other tasks may be aimed more at evaluating performance of proposed designs, or at quantifying resources required to execute a design. In all cases, tasks can be viewed as processes for producing new states of

Function-Orientated Systems

information from existing states. In all cases, both pre-existing and resulting states of information are subject to designers' perceptions of these states, influenced by each designer's particular intentions for a current state.

The problems presented by the example of ground modelling are relevant, perhaps to a varying degree, to all computerised tasks. We are faced once more with our fundamental uncertainty about the completeness of overt knowledge, even when we try to isolate tasks into discrete processes. The concept of well-formedness when applied to tasks can help only those tasks which remain isolated within the bounds of some agreed formal definition. When tasks are intended to serve people who are answerable to other different people, we then have to recognise that function-orientated systems pose the same fundamental issues for computer technology as those posed by integrated design systems.

Influence of older technology

The experience of ground modelling illustrates a prescriptive approach to computing in which a programmer's view of users' intentions is implicitly encapsulated in program code. The computer operations used to model curved surfaces were not accessible to users. The resultant ground surfaces had to be acceptable to designers' further intentions in reshaping the ground, in the larger context of designing other things which would have to be matched to the ground. Eventually, applications included landscaping and earthworks associated with building designs and road layouts.

The curved surfaces generated by the computer program were wholly determined by the coded procedures operating solely on the input data. The effect of the coded procedures was that surfaces were always continuous and smoothly curved, unlike actual ground surfaces, and users had to be satisfied that this difference was immaterial to their purpose in using the modelling program. Problems arose when actual ground exhibited discontinuities, like cliffs or sharp rocky peaks; or when users wished to introduce discontinuities, like sharply bounded flat areas and terracing. Further complexities arose from users' particular requirements for viewing and interacting with surface models to appreciate the form of a surface and impose intended and bounded changes.

These requirements reflected designers' interests, for example, in establishing site lines between different parts of a piece of ground, in locating objects on the ground, and in reshaping the ground to receive the objects. These interests tended to vary for new instances of design projects, and they were never revealed as a complete and general specification for the program. Yet the programmers had to devise complete and bounded procedures which implied a complete specification. This condition was imposed by the use of prescriptive technology and it applied despite the advance from FORTRAN to Unix. Inevitable conflicts between the programmers' prescriptions and the designers' changing interests led to the need for extensive programmer intervention during the useful life of the program.

FUNCTIONS AND INTENTIONS

We began this chapter by stating the general expectation that computers should do useful things, perform useful functions or tasks. The subsequent discussion of experience gives us cause to question this expectation, at least in the context of CAD and especially so in the field of architecture. Do newer developments in software technology offer the promise of a more positive conclusion?

Doing useful things means doing things that people want done. For tasks to be recognised as being valid, people have to recognise value in their products. People have to share the same goals, the same intentions. These observations lead us to consider whether it is possible to describe intentions and to do so in an overt manner which can be represented formally in computers. Of course, it is possible to describe particular outcomes for particular tasks, but the validity of these outcomes depends on the further intentions that people associate with such outcomes. Intentions extend beyond discrete tasks and, in design fields, they appear to be rooted in people's intuitions.

Expert systems

Recognition of people's difficulty in providing overt expressions of tasks they have in mind, in the form of complete step-wise deterministic procedures, has led the field of artificial intelligence to develop expert systems.[5] This development exploits recent advances in logic programming, employing 'true if' or 'if true then' inference rules plus the ability to match incoming expressions with already stored expressions. Rules set out the conditions under which truth values for particular expressions may be identified and, contrary to most previous programming technology, rules may, in principle, be applied non-sequentially. The promise of this rule-based approach is that computers will be able to do useful things with partial knowledge.

The strategy for expert systems is to solicit knowledge from human experts in some defined field, to get these experts to declare what they know about their task, and to get many experts to do so.

Function-Orientated Systems

Knowledge acquired in this way is then constituted as an overt and authoritative knowledge base consisting of facts and rules within a computer system. New instances of task can then be presented to the system, which responds by consulting its knowledge base. Results reflect what a system 'knows', and the system is expected to explain its results by showing what information supports its answers. Cautious advocates of expert systems normally acknowledge that answers produced by these systems should not be accepted as conclusive but, since a system will possess partial knowledge acquired from many human experts, it is also claimed that an expert system can provide better answers than human experts.

As a general condition, expert systems can operate only in precisely bounded fields targeted at well-defined goals. This qualification is necessary to ensure coherence of a knowledge base and orderly operations by a system on its knowledge base. It is also necessary in a more principled sense in order to comply with computational symbolic logic. This condition might be viewed as equally relevant to human experts operating in highly specialised fields, pointing to the significance of the term 'expert' used to label these systems. Attempts to moderate the limitations imposed by this condition, to extend the scope of expert systems, involve proposals for dealing with probability and intuitionistic (or non-classical) logic. Such efforts pose intractable computational problems which remain the subject of speculative research.

In expert systems it is inherent that the knowledge extracted from human experts will be uncertain. These systems are targeted at the intuitive knowledge of experts. They are targeted at that knowledge which people know from their personal experience, without knowing how they know it and without knowing how to provide overt rationalisations to explain it. Knowledge from different human experts, even when targeted at the same goal, is likely to differ and may be contradictory. The contents of a knowledge base will include overt expressions of the outcomes of intuitive knowledge and the knowledge base will, therefore, reflect uncertainty. Thus, these systems can be used only in those specialised fields where people believe that the nature and extent of uncertainty is unavoidable and immaterial to the answers they want from a system. Impressive applications so far include expert systems that fulfil a diagnostic role in matching received symptoms with previously established goals in fields such as medicine and oil exploration. Applications to generative tasks are rare.

We have already discussed design as being dependent on a

combination of overt and intuitive knowledge, and the decisive role of intuition. When expert systems are imported into the field of design, the knowledge they are intended to handle touches upon uncertainties central to decisions affecting design products. We should, therefore, approach expert systems with caution. A loose anticipation of design objects, the lack of prior knowledge of the properties that will describe such objects, cannot be translated into the firm goal specifications required by expert systems. As a consequence, while the outcomes of the intuitions of designers (as broad-based human experts) may be conveyed to an expert system, it does not follow that the system will be able to employ those outcomes as knowledge when dealing with a new instance of design. The role of expert systems in design is inherently limited to discrete analytic sub-tasks of design, and the general reservations described earlier for task-specific systems still apply.

The future of expert systems
The development of expert systems forms part of a larger endeavour to develop and exploit intelligent knowledge-based systems.[6] On this larger front, advances might eventually address some of the issues revealed in the earlier example of ground modelling. The use of logic implemented as generalised inference mechanisms and pattern matching facilities in computers promises a less prescriptive approach to computer applications. This promise rests on the ability to apply rules in a non-sequential manner. It should no longer be necessary to encode complete procedures in which each has a defined start and finish, with a continuous and correct sequence of operations. This advance refers to the encoding of procedures which correspond to a user's view of a task; the purpose of the coding is to gain access to a computer's internal procedures, the lower level programs, which eventually rest on binary operations. By having a machine implementation of a general logic environment, applications can be formulated as logical expressions and the necessary mappings to machine operations can then be executed automatically by the system. Within the limits of the logic environment, it should, in principle, be possible to redefine or extend a computer application without calling for programmer intervention.

If this promise can be fulfilled, then we should no longer be dependent on complete correspondence between a user's perception of a whole task and a programmed representation of the same task in a computer. The computer should be able to work with partial knowledge and accompany its answers with explanations

Function-Orientated Systems 131

which will prompt the user to supply further knowledge. The computer's knowledge should grow to match the knowledge of its user.

We are now identifying ambitions that are shared across the whole field of artificial intelligence and they are widely recognised as embracing fundamental problems that cannot, as yet, be resolved. As an example, no one knows how to deal with the unintended effects of adding further rules to a growing knowledge base — any imposed control procedures would diminish generality and increase prescriptiveness. Such issues will be discussed further in Chapter 7.

Beyond expert systems
Even if these larger ambitions can be fulfilled, certain fundamental questions will remain. We need to be clear about what we mean when we refer to computer programming languages and logic environments, and the advantages associated with a generalised logic environment. Just as a computer programmer needs to know the environment provided by a programming language in order to be able to encode a user's application in the language, so a logic programmer needs to know the environment provided by a system of formal logic. If computer users are to formulate their own logical representations of their own tasks, they will need to be fully familiar with the logic environment offered by their computers. Working in a formal logic might prove to be more difficult than working in a programming language. If we then require specialist logic programmers to be interposed between users and their computers we will be confronted once more by all the issues of older computer technology and we will lose our sense of achievement.

For logic to be usable by ordinary people, these people will need to have interpretations between logic and their own common-sense experience of our world. Whether, alternatively, such interpretations can become the responsibility of the formal environment which receives informal expressions from people remains an open question. Research into natural language understanding can be regarded as taking us towards this goal. If computers can be made to understand people's ordinary use of language, then people might be able to tell computers what they want them to know and do.

Even if people could converse with computers, and if computers could answer back in a manner that would elicit further knowledge, and if they could then maintain coherent logical representations of whatever users tell them, we would still be left with a problem. If we say things to computers with the expectation that they will respond by doing things, performing useful tasks, and if we expect their performance to match our perceptions of tasks, then computers will need to know something about human intentionality.[7] As we have seen in the example of ground modelling, formulation of a task with respect to states of information before and after a task is executed is conditioned by the intended purpose for the result in some larger context of people's activity. The issue can be put quite simply. In general, when you ask something, the answer will be conditioned by knowing why you are asking. It is intentionality that gives meaning to the words you use. If you ask things of computers, the usefulness of answers will similarly be conditioned by the computer 'knowing' why you are asking. Does this mean that the usefulness of computers is dependent on computers first knowing all about people; and does such knowledge have to be agreed among people?

The problems I have indicated here arise from the idea that computers can be made to do useful things in the manner in which people do things, and that they can use human language to do so. This now seems to be an implausible ambition. My discussion on language and knowledge in Chapter 2 draws a distinction between knowledge including logic within people, as human knowledge, and externalisations of knowledge, as expressions and formal logic. There I suggested that people's use of formal logic on their expressions, to make expressions describe useful behaviour, ought to be regarded more like people's intelligent use of dumb machines. I will develop this position in Chapter 7, and I will generalise it into a discussion of computer literacy in Chapter 8.

WHY WE HAVE TO BOTHER

Why should we bother about all these qualifications? Surely computers exist already that perform useful tasks and might we not expect gradual and cumulative progress towards more generally useful systems? Relevant to the field of architecture, we already have environmental simulation programs which calculate energy requirements, lighting distributions, and acoustic performance, and we have computer models which support structural calculations.

It will be instructive to look once more at the current position. For each of these tasks we have many programs, each employing slightly different assumptions and each producing different answers. This situation is nicely illustrated by a study of computer programs for analysing and designing reinforced concrete continuous beams, which noted more than twenty programs and carried out a comparative evaluation of seven programs.[8] Each complied with the new (1972) British Standard Code CP110. The conclusions of the study contain a noteworthy paragraph:

> We evaluated seven programs all of which have the same aim; the analysis and design of continuous beams according to CP110. We tested these programs on identical problems. The tests were carried out by the same team on all programs, and every effort has been made to give every program the same information. Yet, as the reader will have observed from preceding sections of this report, we obtained very different results from each program. These differences in results cannot be explained by deviations from the code which is adhered to by all programs with only insignificant exceptions. The code, however, does not consist of a definitive set of rules. It allows the engineer, as a code should, a wide margin within which to apply judgement and take decisions. It is these decisions the programmer builds into his program, and it is his judgement that is reflected in the differences of result; . . .

The study disclaimed any intention of presenting a formula for selecting the best program, but it did express concern over duplication of programs. Discussions at conferences following the study, including participation by the Department of the Environment which commissioned the work, were more pointed in expressing concern about the multiplicity of programs and the divergence of results. There was a popular belief that it would be better to have just one correct program.

The interesting point of this example is that, variously, all the programs were correct in so far as the Code provided a measure of correctness. We can then interpret this position as meaning that the programs represented various legitimate states of knowledge pointing to an evolution of knowledge among a collective body of engineers (reflected by their programmers). Taking the view of design discussed previously, this evolution is essential to the practices of engineers. If, instead, we were to encapsulate a consensus of all engineering knowledge in one program, we would inhibit further evolution of that knowledge.

Existing function-orientated systems present 'snapshot' views of states of knowledge about selected aspects of design. They do so in a manner that does not readily accommodate changes in knowledge, in application fields where people's knowledge evolves. New knowledge prompts new programs, leading to a multiplicity of programs nominally performing the same task. We should not accept any such program as being conclusive, as properly representing some aspect of design, and as conditioning future design practice.

Furthermore, there is the question of who can use these programs. The varied assumptions built into programs have the effect that each program employs its own unique model and requires input procedures from users. As a consequence, these programs tend to be most useful to those organisations that develop them. Other potential users find difficulty in understanding the models or do not agree with the assumptions built into the programs. In the field of architecture, we can observe that applications appear most successful only in the offices of specialists whom architects consult. Few function-orientated systems are in regular use within architects' offices.

Computer programs that presently do well in architects' offices are those which do not impinge on design knowledge, which do not perform tasks specific to the practice of architecture. Examples of programs that do well are word-processors, plus extensions which serve administrative tasks, and drawing systems. These applications will be discussed in the next chapter. If we expect to do more, if we really want to employ computers more usefully, then we require fundamental advances in our understanding of all the issues identified in this chapter.

SUMMARY
In this chapter we have looked at the experience of developing a ground modelling program as an example of a function-orientated system, and we have looked at the development of expert systems. Observations have been directed at the general expectation that computers should do useful things; they should perform functions or tasks that people can recognise as being valid.

Function-orientated systems perform discrete tasks in precisely bounded fields, targeted as well-defined goals. These conditions simplify computational problems — task specific programs are easier to program. However, these same conditions pose severe problems for program users. People have to agree on precise formulation of tasks, on the boundedness of these tasks, and on the validity of their results.

When computerised tasks are intended to serve people who are answerable to other, different people, we find ourselves faced with intractable tasks, and we then have to recognise that function-orientated systems pose the same fundamental issues as those posed by integrated design systems. We have to deal with uncertainty and with unforeseen evolution of knowledge among people. The ambition of getting computers to do useful things presents us with profound problems which we do not yet fully understand.

Function-Orientated Systems

Expert systems represent an advance in technology, but we will need to progress beyond a position in which specialist logic programmers replace more conventional computer programmers. We might expect a generalised logic environment to include interpretive processes for dealing with people's informal expressions of their common-sense experience. Thus, we come to speculative ambitions for natural language understanding. Thereafter, we will still be faced with the problem of getting computers to know people, to accommodate human intentionality. Meanwhile, practical computer applications are limited by our use of prescriptive technology. Computer programs that presently do well within architects' offices are those which do not impinge on design knowledge. This explains architects' use of simple word-processors and drawing systems.

Notes to Chapter 5
1. More examples of task-specific systems are described by Eastman (1975) and Mitchell (1977).
2. This task-specific philosophy is described further by Maver (1977).
3. The issue of quantitative versus qualitative tasks is discussed further by Bijl (1979b).
4. This ground modelling application is described by Bijl *et al.* (1980) and it arose out of early work at EdCAAD on modelling strategies supported by the UK Science and Engineering Research Council.
5. Strong advocacy for expert systems has come from Michie (1979), and their use in the construction industry is described by Lansdown (1982).
6. The present popularity of intelligent knowledge-based systems and AI in general was stimulated by the Japanese announcement of its Fifth-Generation Computer Systems project, described in JIPDEC (1981) and by Fuchi (1981) — the UK response was Alvey (1982).
7. Intentionality, referring to a causal link between mental states and actions, is a complex topic which is outlined in a readily understandable way by Searle (1984) in his chapter 'The Structure of Action'.
8. This study of different programs performing the same task is presented in a Construction Industry Computer Association Report by Bensasson (1978).

CHAPTER SIX

DRAWING SYSTEMS

My drawings depict what I know in a manner that words cannot explain—how, then, can I tell a computer to produce my drawings?: we will now look at drawing as a mode of expression, and the use of computerised drawing systems.

Following the experience of integrated CAD systems during the 1970s, including the kind of difficulties already discussed in Chapter 4, CAD developments in the late 1970s and 1980s were far less ambitious. These systems enable people to use computers to produce drawings; the drawings are the intended end products of using a computer. The key difference from earlier integrated CAD systems is that in the case of drawing systems the computer knows nothing about what is being depicted. The computer knows only about lines and points, and about edits and transformations which can be applied to collections of lines and points. The purpose of these systems is to produce drawings faster and of a better quality than might otherwise be produced by hand.

Lest this general characterisation of drawing systems be challenged as being too exclusive, it should be added that some drawing systems do have limited abilities to handle other information associated with drawings. A typical example is the use of solid geometry modelling to support drawing projections of three-dimensional objects. Another example is the provision of schedules by means of quantifying instances of parts of a drawing. We will comment on such cases later; meanwhile, it will be useful to proceed with the strict characterisation of dumb drawing systems.

We can start this discussion by drawing a close parallel between drawing systems and word-processors. Word-processors know about possible arrangements of characters, and about edits and certain transformations which can be applied to collections of

Drawing Systems

words, without knowing what the words mean. Some word-processor will also provide associated functions, like a spelling checker, for example. Where a word-processor offers a screen editor for text formatting, its similarity with drawing systems is apparent in the essentially graphical nature of its screen operations.

This similarity between drawing systems and word-processors is significant. Apart from certain specialist fields of computer application, word-processors are the one widely recognised and successful application of computer technology. This success rest on the fact that word-processors are dumb. Since these systems do not know the meanings of any words they process, such knowledge cannot conflict with different knowledge intended by any user. Yet, since there are regularities in the ways in which all people operate with words, irrespective of the different things people intend to say with words, word-processors can be made to exploit these regularities and, thus, they can be useful to many people. Coupled with the trend towards lower cost and higher power computers, this charactisation of word-processors explains why they can be targeted at mass markets.

The idea for drawing systems, then, is that they should follow in the path of word-processors. To do so, there have to be similar regularities in the ways in which all people who draw operate with drawings. In this chapter we will explore the concepts that underlie drawing systems, and we will consider their implications for designers.

DUMB SYSTEMS

What does being dumb mean in terms of properties of a system? Dumbness refers to the distinction between a representation environment and things that are represented in that environment, as set out in Chapter 2. Word-processors present an environment of characters in which words can be represented. This environment has no responsibility for interpretations that will condition instances of words. The definition of the environment, plus the operations that can be applied to it, are not conditioned by any anticipation of particular words or word compositions. Yet the environment is useful for constructing any words and word compositions.

The virtue of being dumb is that any operations invoked in the environment are not conditioned by any need to preserve an intended model a user might have a mind. Within the confines of the environment there are no correct or incorrect operations. Any such notions are matters for the user and his or her interpretations with respect to a world outside the environment. A word-processor cannot be wrong, but any construction of words might be right or wrong as seen from outside the word-processor.

In the case of dumb drawing systems we want to achieve a similar distinction between a general drawing environment and drawings which correspond to particular models users might have in mind. For this purpose, we should expect the environment to be defined in terms of lines and operations on assemblies of lines. It should exclude any further higher level constructs which are specific to certain models, which might condition the environment by the need to preserve a particular model. If that were to happen, the drawing system would lose its generality and fewer people would be able to use it. We will now consider whether this ambition is feasible.

Words and drawings
First, let us consider briefly what kinds of things words and drawings are, and how we use them. Are they similar? Earlier we discussed language and, in particular, written expressions and their interpretations in human knowledge. We need to restate here only that written words are objects which can be read as symbols which stand for things in some knowledge domain. The key point is that words denote symbols and our use of symbolic constructs determines the way in which we use words in arrangements of word compositions.

Symbolic objects
Written words, made up of characters, can be thought of as drawn objects. However, the role of words in depicting symbols has the effect that it does not matter too much how they are drawn. The depiction of a symbol does not have to look like the thing that the symbol represents. Quite a lot of distortion is permitted before the symbol depicted by written word is changed or lost. Note that this tolerance is allowed mainly for individual characters, and less so for arrangements of characters within words, and arrangements of words in word compositions.

This tolerance is evident is the range of different character fonts used for text (Figure 6.1) and in the variety of people's handwriting. Our ability to read will also tolerate some misspelling and even incorrect grammar. Indeed, such variations may be interpreted as intended expressions of intonation or emphasis which might add to the symbolic content of otherwise regularly formed (or drawn) words. The more we attach significance to the actual depiction of words, so the word objects become pictorial objects.

This transition from symbolic object to pictorial object can be

Figure 6.1 *Written and drawn expressions*. To write the word *dog* I assemble a string of characters that looks like a symbol for dog — it does not have to look like a dog. If I draw a dog, then the drawing must look like what I think a dog looks like — it must depict spatial properties that describe a dog. Note that the depiction does not have an obvious decomposition, but I must have control of all its parts to differentiate a dog from other things, such as a cat.

observed through a range of drawing objects from word characters, through ideographs and diagrams, to verisimilar or pictorial drawings. Figure 6.2 presents such a range. What distinguishes a pictorial object from a symbolic object?

Analogic objects
A verisimilar drawing is one in which the spatial properties of elements in the drawing correspond to spatial properties of some other object which it depicts. The drawing provides a visual semblance of some perception of actuality, be it a perception of something concrete or abstract. We can say, therefore, that a verisimilar drawing serves as an analogue for certain properties of the thing it depicts. To put this point more strongly, from the point of view of the person who makes the drawing, a verisimilar drawing *is* the thing it depicts in the sense that it is what the person knows about the spatial properties of the depicted thing. In this sense, a verisimilar drawing is the opposite of a symbolic object.

The definition of verisimilar or pictorial drawings has implications for the ways in which pictures can be composed. The form of a drawing, its composition of lines and the spatial relationships exhibited in these compositions, becomes vitally important to the job of expressing what its originator has in mind. This close connection between knowledge within people and expressions of such knowledge should lead us to expect that the process of drawing will be idiosyncratic to each originator. The effect of this position, in comparison with our use of words, is that drawings generally are not assembled from predefined discrete pieces of drawing. For drawings, we do not have things that correspond to established characters or words which can be assembled into compositions. Instead, the process of drawing requires access to the kind of primitives that are used to form characters and to attach one character to another. The process of drawing requires access and control of the lowest level entities that contribute to the visible presence of drawings—this position will be discussed further in Chapter 7.

Design drawings
It might be argued that the preceding paragraphs refer only to the extreme case of verisimilar drawings, and are appropriate to drawings as an art form and not to practical drawings of the kind produced by architects. We can try to moderate our definition by saying that verisimilar drawings contain arrangements of lines whose properties depict known properties of other things. This

Drawing Systems

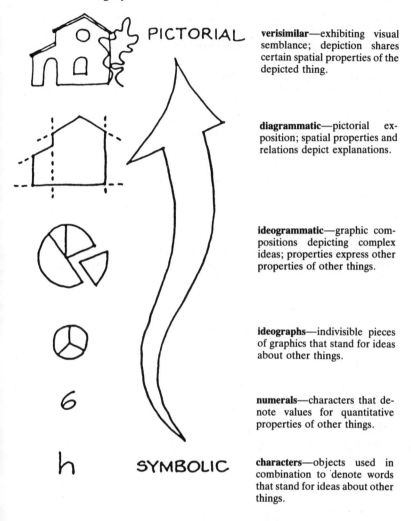

verisimilar—exhibiting visual semblance; depiction shares certain spatial properties of the depicted thing.

diagrammatic—pictorial exposition; spatial properties and relations depict explanations.

ideogrammatic—graphic compositions depicting complex ideas; properties express other properties of other things.

ideographs—indivisible pieces of graphics that stand for ideas about other things.

numerals—characters that denote values for quantitative properties of other things.

characters—objects used in combination to denote words that stand for ideas about other things.

Figure 6.2 *Range of drawing objects*. Objects are differentiatied by their roles in depicting knowledge. These roles range from objects which denote symbols that stand for certain meanings in particular knowledge domains, to objects which serve as analogues that share spatial properties of other things they depict. This difference between symbolic and analogic objects affects the way in which they can be produced and read.

introduces an assumption that spatial properties of other things can be known in the same way to both the originators and the readers of a drawing, which should bias drawings towards realistic depictions of known concrete objects. The analogic role of verisimilar drawings persists, but the range of application is restricted.

We should then consider that the concrete objects drawn by designers do not yet exist. Design drawings depict propositions about future objects. It follows that descriptions of those objects cannot be known to other people before they are designed, otherwise we would reduce design to a mere process of selection. Thus, it remains a possibility that the assumption required for the moderated definition of verisimilar drawings may not be fulfilled. This likelihood is affected by the degree of innovation included in a design, and by the extent of shared experience between the designer and other people at whom the design is targeted.

A further condition on designers is that their designs should be capable of being executed, usually by other people. In the case of architects, drawings have to serve as instructions that can be understood by builders. Architects have to produce diagrammatic drawings, in terms of the range shown in Figure 6.2, including symbolic objects which rely on shared conventions. The techniques used to produce and read these drawings can be described overtly and can be learned, so we might expect that the process of drawing will begin to exhibit regularities. Here we are coming closer to the conditions that explain the success of word-processors.

With these initial observations on the comparison between words and drawings, we should begin to expect that systematic approaches to drawing operations will result in systems that differ from normal drawing practice. The question is whether these differences are important to people's ability to express themselves by means of drawings.

Ordinary Drawing Practice

We will now review conventional drawing practice in order to establish a basis from which to judge computerised drawing systems. Examples will refer to architectural practice, but they should also be viewed as relevant to other fields of design. Emphasis will vary in different fields, as, perhaps, in the case of increased emphasis on precision and reduced emphasis on innovation in mechanical engineering. As said before, architecture highlights issues central to design synthesis in all fields.

Drawings convey spatial information, providing descriptions of spatial relationships of depicted objects. By adding annotations in the form of text or other symbols, non-spatial properties can be associated with depictions. We will restrict our discussion to line drawings and drawing objects which are conditioned by higher level knowledge of abstract things like geometric objects — excluding wider consideration of visualisation graphics and image processing.

Role of drawings
The role of drawings is to convey perceptions in the minds of designers in an externalised form that representations of can be seen by other people. Here we are considering externalised representations of knowledge. However, we do not have overt and widely agreed conventions which govern both the production of all drawing objects and the unambiguous interpretation of these objects. We do have formal understanding of certain objects, such as our knowledge of geometry, but this knowledge does not condition all drawing 'utterances'. It seems that drawings also depict intuitive knowledge about other things that are drawn — drawings are evocative expressions.

When drawings are intended to convey instructions, they are required to convey logical coherence only to the extent that their spatial information should be repeatable outside a drawing, in some other medium. In general, there is nothing in a drawing object itself which can tell us whether a drawing is correct with respect to non-spatial information.

The evocative nature of drawings, as applied to spatial information and associated with non-spatial information, makes drawings a powerful mode of expression for designers. We then call upon formal knowledge by talking about drawings, by attaching annotations.

Architects produce drawings to convey information about designs to other people and to themselves. Drawings may be impressionistic, to show what a building will look like, to visualise the design. Such impressionistic drawings commonly employ three-dimensional projections (Figure 6.3). Drawings may be diagrammatic, to show the organisation of spaces and the arrangement of building fabric. These diagrammatic drawings usually employ orthogonal projections, plans, sections, and elevations prepared for clients (Figure 6.4). Drawings may be instructive, telling other people how to execute the design, build the building. Instructive drawings, or working drawings, are usually an elaboration of diagrammatic drawings, combining graphic, numeric, and text expressions to convey precise and unambiguous information (Figure 6.5).

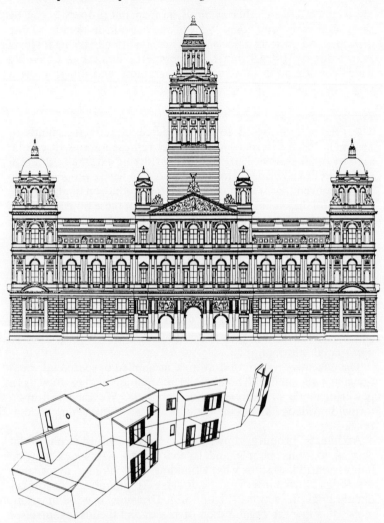

Figure 6.3 *Visualisation drawings*. Visualisation drawings include polished presentations of orthogonal projections. Perspective projections are used to give an impression of the three-dimensional properties of design objects during the course of being designed. Usually, perspectives have to be constructed from orthogonal drawings that define objects, using procedures for collecting data from these drawings in order to position perspective lines in two-dimensional space.

Figure by permission of Compass Graphics and DeCAL.

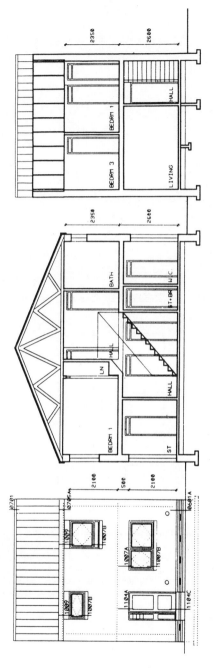

Figure 6.4 *Diagrammatic drawings*. Diagrams are intended to convey unambiguous information on spatial arrangement, employing the conventions of orthogonal projections. Architects employ these conventions to convey plan, section, and elevation views of their design objects, buildings. Other people reading these drawings need to share these conventions.

Figure by permission of the Scottish Special Housing Association.

Computer Discipline and Design Practice 146

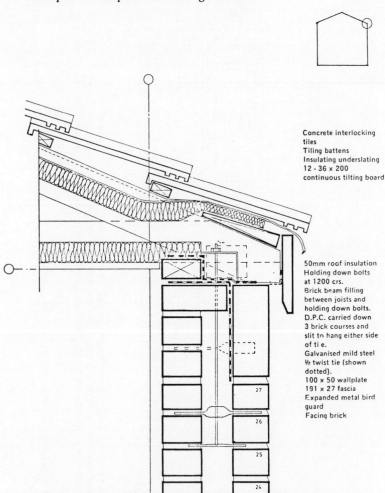

Figure 6.5 *Working Drawings*. Diagrammatic drawings intended to convey instructions for executing a design, constructing a building, are more dependent on text annotations and metric dimensions. Interpretation of annotations written by architects on their drawings, with respect to parts of buildings depicted by lines in drawings, requires knowledge both about drawing conventions and building practice. This knowledge has to be shared by other people who read the drawings.

Figure by permission of the Scottish Special Housing Association.

Familiarity with traditional drawing operations means that an architect can reveal what he or she has in mind by letting drawings pour from his or her fingertips. The act of drawing does not require a conscious effort separate from thinking about the things being drawn. The act of drawing is the means by which architects manifest their contribution to a design, thereby making their presence as designers evident to other people. Designing and drawing appear to be inseparable.

The importance of this link between thinking and drawing is evident when architects draw in order to reveal to themselves unanticipated properties of other depicted objects. This can occur, for example, when working out a junction detail between two or more pieces of building fabric. Drawings are not produced at the end of a completed process of designing, they are part of that process. This remains true even for those drawings that serve as instructions for executing designs.

Traditional drawing technology
The traditional technology for producing drawings includes a two-dimensional flat plane providing the drawing space (paper attached to a board), a pointed device for marking the drawing space (pencil or pen), a device for deleting marks (a rubber), plus straight-edged devices for producing lines (tee-square and set-square), and a device for drawing arcs (a compass). Less generally, further devices serve as templates for curved lines (french curves), circles, and other simple pictorial forms. This equipment, illustrated in Figure 6.6, defines a drawing system in which all drawing objects are two-dimensional and are composed of lines. The system induces a bias in favour of straight lines and, in particular, vertical and horizontal lines.

This technology might not be the product of wilful invention. We can note that the natural law of gravity favours vertical and horizontal lines. Once we have a technology for producing drawings, however it was originated, that technology has an influence on what we draw and what we can read from drawings. The technology influences what we can design. In general, designers cannot design things that they are unable to depict in drawings.

Conditioned by traditional drawing technology, we can regard drawings as compositions of lines. These lines are produced by moving a marker along a straight-edged or curved template. The position in which a line is to be placed in the drawing space has to be decided with respect to other lines that already exist in this space, and this is usually described by drawing lines through existing lines

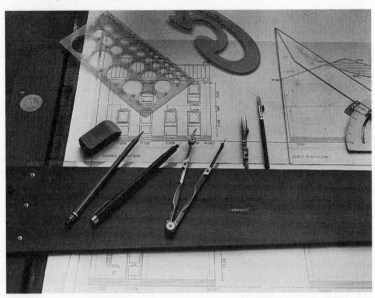

Figure 6.6 *Traditional drawing technology*. Ordinary drawing practice employs the technology defined by pencils and pens, rubbers, drawing boards, tee-squares and set-squares, compasses and curved line templates. These items define a line drawing system that can operate only in a two-dimensional drawing space. The system induces a bias in favour of straight horizontal and vertical lines. As with any system, in using this system we have to realise things we have in mind in things that can be produced by the system.

and using line intersections to locate or delimit further lines. A composition then employs a structured classification of lines differentiated according to their role in the drawing. Thus, we have construction lines for purposes of setting up line intersections (light pencil lines), finished line segments that depict shape properties of depicted objects (ink lines), dimension lines (like construction lines, but with annotations for length values), and the boundary lines of the drawing space.

Lines are the primitive drawing objects. People who draw in this environment learn to equate lines with the spatial properties of things they have in mind when they draw — people draw things not drawings. As the designer's knowledge of things changes, so he or she changes the lines depicting that knowledge in the drawing.

This use of lines is possible because the person who is drawing has full control of the lines. Traditional drawing technology does not impose any higher level interpretation on assemblies of lines such

Drawing Systems

that particular relationships have to be maintained when a drawing is changed. Edits to line drawings simply involve the addition and deletion of lines to change, for example, the locations of line intersections and to reposition line segments. Higher level constraints, such as might come from knowledge about geometry, are called into play only if a person sees them as part of the description of something that is being modelled in a drawing and, even then, they do not become part of the drawing.

If, for example, we see that an architect has drawn a rectangle and calls it a door, he or she will think of the composition of lines as being the door. If the architect wants to change the height of the door, he or she might reposition the top of the door upwards, and extend the two sides to meet the new top of the door, if the door is intended to remain rectangular. However, there is nothing in the drawing to prevent the architect from raising only one end of the top of the door and extending only one of its sides, to produce a door with a sloping top. This might upset our recognition of the rectangle we see as describing the door, but the geometric property of rectangularity is not part of the drawing and it might play no part in the architect's perception of the door.

Interpreting drawings

Bear in mind that so far we have been considering lines only as objects that occur in drawings and not as objects that might occur in other environments, such as abstract lines in Euclidean geometry. Such other lines may be depicted by drawing lines. Knowledge about the other lines can then be used to decide where drawing lines are to be positioned and may also be used to interpret line drawings.

Interpretation of drawings refers to the act of taking information contained within a drawing and recasting it into another form that can become part of another description in some other knowledge domain. So far, we have claimed that a person who is fully familiar with drawing does not differentiate between interpretation and the act of drawing. This may also be true for similar people when they read each other's drawings. However, when dissimilar people have to read these drawings, and if they have to receive explanations or instructions from the drawings, then interpretation does become a deliberate and separate act.

Interpretation depends on recognition of instances of drawings as models of other things, plus knowledge of those other things being modelled. Such recognition is not part of the drawing system and is the responsibility of people applying their knowledge to instances of models — much in the way in which people are responsible for words that may be produced by a word-processor. Drawings might invoke knowledge in people about such things as drawing projections and about depicted objects such as buildings.

The interesting point is that traditional drawing technology does not pre-empt any of this knowledge. The technology is aimed purely at the production of lines and compositions of lines. Thus there is no (or very little) likelihood of conflict between the technology and any person's interpretations. There might be conflict between something a person has in mind and what he or she sees in a drawing, but that would be a conflict between the person who uses the drawing environment and another person who reads a drawing produced in that environment. Such conflict typically prompts further dialogue between those people.

Drawings as models

Any communication by means of drawings passing between designers and other people entails some form of dialogue between them. People appear to be good at conducting a dialogue. They point at blank spaces in a drawing and are able to say what they mean so that lines which describe boundaries to spaces can be identified; and they go on to draw or talk about changes to things depicted in the drawing. Ambiguities get recognised and resolved, perhaps, by the ability of people to call on any further knowledge they might have in mind.

If people employ formal models to support their interpretations, such models are not essential to traditional drawing technology and they do not need to be preserved during drawing operations. A person who is targeting a drawing at a model, does not need to ensure that the drawing corresponds to the model while it is being altered, but he or she has to ensure only that the model is restored when the drawing is once more presented to someone else.

Consider the example shown in Figure 6.7. We have a drawing of a house with a porch covered by a gable roof, and it is proposed that the porch be made narrower. The drawing is an object consisting of lines defined by angle values and by length values between line intersections. We can draw the proposed change in various ways. Line intersections can be moved so that line angle and segment length values change, or we can move line intersections while maintaining line angles. Each case produces different effects. We can then proceed by various successive steps to achieve the same eventual result. If we know that the porch is a rectangle and the gable is a triangle, we can describe the change by specifying transformations to these geometric objects, such as scaling each with respect to their origin, assuming that someone knows about the

Drawing Systems

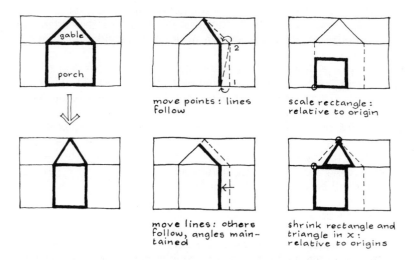

Figure 6.7 *Modelling with drawings*. The first drawing (upper left) describes a house with a porch and its gable roof, and the one below (lower left) shows the intended change. The middle two drawings show the change being made by moving line intersections, with their line angle values being either flexible or fixed. The last two drawings (upper and lower right) show the change being specified as transformations applied to higher level geometric entities. Together, these examples illustrate alternative modelling strategies for arriving at the same result.

transformations and is able to execute them. Again we have different effects, and again we can define a succession of transformations by which we can proceed to the same result.

This example illustrates a number of points about ordinary drawing practice. First, if we employ a higher level model to help our interpretation of a drawing (as in the case of a rectangle which is depicted by lines and which describes the porch), it is not necessary to preserve the model (the rectangle) while the drawing is being changed (to depict a narrower porch). Secondly, if we describe the change to the drawing by specifying a transformation of a higher level object (the rectangle), it does not necessarily follow that the person specifying the change can be more certain of the outcome. Thirdly, any outcome remains valid as a drawing in the sense that the spatial information depicted in the drawing can be represented outside the drawing. People reading the drawing have to decide whether it is acceptable by deciding whether the drawing can be interpreted as something else they have in mind. This last point is interesting because it restates the position (expressed earlier in

Chapter 2) that drawing environments cannot include criteria for the correctness of their drawings. Correctness is dependent on knowledge of objects as known in domains outside the drawing system.

Drawings themselves can be regarded as models, but their role in depicting spatial information makes them behave more obviously like physical models. They can be regarded as behaving in much the same way as plasticine models. These cases are characterised by the lack of formalised conventions for linking particular models to any other things that they might model. This same characterisation also applies to word-processors and it is this that distinguishes word-processors from language understanding systems. This general point will be developed further in Chapter 7.

Things that architects draw

To round off this discussion of ordinary drawing practices, I will now consider some of the characteristic features of architects' drawing practices. First, in common with drawing practices in many other fields, there is a general rule which says that any drawing should be started by first setting up its outlines and then the drawing should be developed gradually, by filling in the details. This can be called the top-down approach to drawing, as shown in Figure 6.8. Thus, drawings of buildings are started by setting up the controlling construction lines to describe their main constituents, and these are then subdivided by adding further lines to describe lesser constituents. This process continues until, in the case of building plans, the drawing depicts individual walls with doors and windows. Construction lines commonly appear as light pencil lines which can be erased after final details have been confirmed. Initial construction lines have to be positioned somewhere within a drawing space, on a blank piece of paper. It is quite common for the drawing space to be covered with a regular pattern of construction lines in the form of a grid used to position further construction lines and finished lines.

Secondly, the geometry of most buildings is mostly orthogonal, and most components of the fabric of buildings take the form of flat slabs or point supports, such as walls, floors, and columns (Figure 6.4). Thus, drawings of buildings largely depict orthogonal spatial relationships and they generally include many pairs of parallel lines. Walls drawn in plan and section, for example, will appear as parallel lines separated by a constant space, in a similar way for many instances in different buildings by different architects.

Of course, buildings can and do also exhibit a great variety of

Drawing Systems

other spatial relationships. These tend to occur at the junctions between components and in the details of components (Figure 6.5). It should be noted that the occurrence of these more varied spatial relationships is not necessarily conditioned by prior geometric knowledge. Such knowledge might be applied only later when these relationships are refined to produce final designs.

Thirdly, as single objects, whole buildings are compositions of many other objects (Figure 6.9). These are combined to form the fabric of buildings and define the spaces within buildings. The

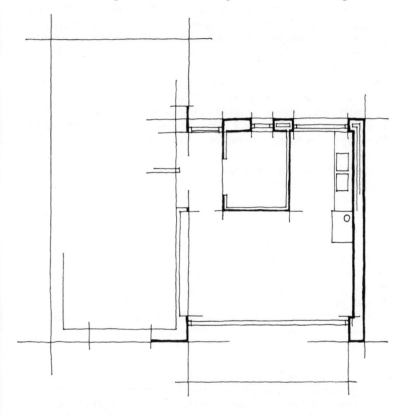

Figure 6.8 *Constructing drawings*. A drawing is usually constructed by first putting down the main construction lines, or outlines, and then adding further lines to subdivide existing lines, until the required details become apparent. This is normally called a top-down approach and it is generally regarded as good drawing practice. As a development of this approach, construction lines can be regularised in a grid pattern over the whole drawing space. Grid values then assist in establishing dimensional precision when a person draws or reads further lines.

composite spatial description of a whole building can be very complex, made up from varied combinations of many relatively simple parts. We can note here that this characterisation of whole buildings excludes buildings from most of those advances in computational solid geometry modelling that are aimed at singular complex objects.

Fourthly, the same parts can occur many times in a building, and in different buildings. This leads to the practice of differentiating between general arrangement drawings and detail drawings, in which the former provides locational information for the latter (Figure 6.10). However, the sameness of parts is generally not complete. A component or detail generally cannot be described within the boundaries of its own separate drawing so that its description remains unaffected by any context in which it may be used. Repeated use of the same part tends to entail modifications to instances; detail drawings get altered.

Lastly, annotations are added to drawings to say what the things are that the drawings depict. Thus, the label 'kitchen' attached to a drawing will indicate that the lines around the label provide a spatial description of something that we otherwise know to be a kitchen. Dimensions are an important kind of annotation used to attach metric values to lines. As a general rule, written dimensions always take precedence over dimension values that might be measured directly from the drawing. Thus, for example, a plan view of walls drawn as parallel lines may show a constant width for walls of different thickness. Similarly, dimensions for rooms might be given different values, depending on whether the lines depicting the walls are understood as referring to finished surfaces or the unfinished primary fabric (Figure 6.11). This practice has the effect that drawings do not need to consist of precisely positioned arrangements of lines. However, it also follows that correct readings of drawings are dependent on the drawings being directed at people who already know a lot about buildings.

Computerised Drawing Systems

The previous section has described ordinary drawing practice: traditional drawing systems. Implicitly, this description contains a specification for computerised drawing systems. Ideally, the definition of a system should be independent of its implementation and should survive implementation within computers. This turns out to be a surprisingly difficult ambition.

As with word-processors, we want drawing systems that are useful for drawing anything. A drawing system should offer a

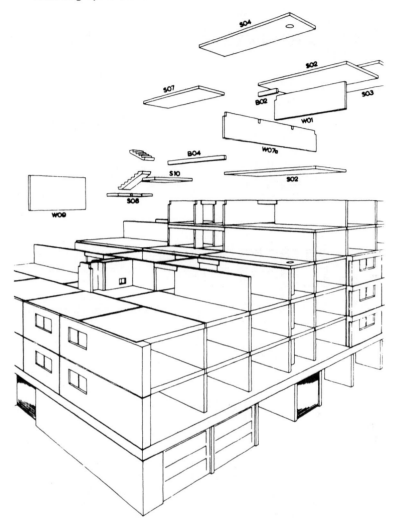

Figure 6.9 *Compositions and decompositions*. Drawing parts have to correspond to some decomposition of a whole perception of an object which the drawing depicts, a building. In this example, the decomposition follows the factory produced components that were used in the assembly of the building (Bijl, 1968). More usually, decompositions are not so clearly defined and they can vary for the same object according to the different perceptions of the people who draw them.

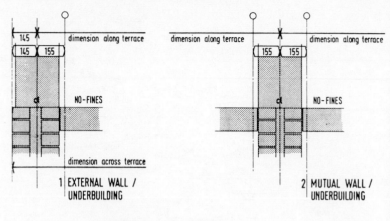

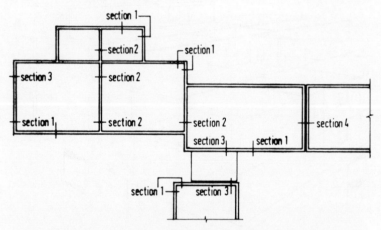

Figure 6.10 *Locating details*. The same parts can occur many times in a building, or in different buildings. This leads to the usual practice of differentiating between general arrangement drawings and detail drawings. The former provides locational information for instances of the latter. To achieve satisfactory compositions of details, parts of detail drawings must have known mappings to parts of general arrangement drawings.

Figure by permission of the Scottish Special Housing Association.

Drawing Systems

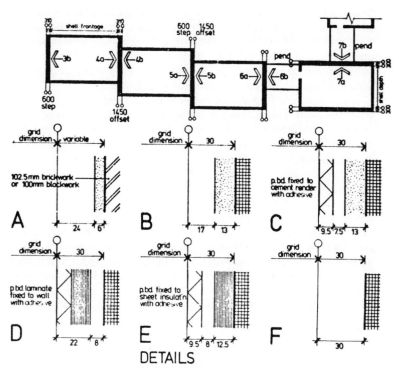

DETAILS

Figure 6.11 *Dimensioning*. Dimensional precision usually does not rely on the precision of a drawing but, instead, relies on knowledge about objects depicted in the drawing. If a part of a drawing is known to depict a wall, and the part is identified by reference to a dimensional grid, then dimensions can be calculated to the nonimal grid line, to the unfinished surface of its primary material, or to the outer face of the finished wall. These calculations require knowledge of the details of the wall construction.

Figure by permission of the Scottish Special Housing Association.

general environment for constructing drawings which users can read as spatial models of depicted things. The modelling environment should consist only of entities and relationships necessary for constructing and manipulating arrangements of lines, just as a word-processor offers entities and relationships necessary for constructing and manipulating arrangements of characters.

We might expect a drawing system to be defined in terms of line entities with different type and style values, angle and length values, and compositional properties of intersection and attachment. The system should include operations for editing drawings, by adding or deleting lines, and operations for transforming arbitrary selections of line values. Superficially, these do not appear to be severe requirements. However, problems emerge from our general lack of understanding about drawing environments separated from particular intentions for drawings. Systems become particular and they then frustrate the separate intentions of their users.

All computerised drawing systems exhibit certain similar characteristics. In the following paragraphs we will explore their common characteristics and consider how they relate to people's normal expectations for drawings. Again, examples will refer to architects' use of systems.

The act of drawing
When using a computerised drawing system, a computer is interposed between an architect and his or her drawing. The computer executes the drawing process, and the architect has to tell the computer what he or she wants it to draw. The computer has to know enough about drawings to enable it to understand what drawing operations it is being asked to execute. Fundamental to drawing systems is the idea that the computer should not also need to know anything about other things that might be depicted in drawings. It does not have to know what it is drawing.

Whole drawing operations, as perceived by architects, have to be decomposed into sub-tasks which can be represented as computer operations controlled by a computer program. The architect then controls drawing operations by selecting instructions that are included as parts of the program. The architect has to think of the act of drawing as a process of construction that can employ only the available computer operations (Figure 6.12). He or she has to separate the act of drawing from thinking about what is being drawn. This position can be compared with that in which an architect employs a technician to do the drawing, where the architect cannot handle the pencil and the technician knows nothing about buildings.

Drawing systems tend to be biased towards decompositions of drawings that can correspond to well-bounded computer operations. They favour drawings that consist of assemblies of well-defined and discrete drawing parts. Thus, most drawing systems see the act of drawing as a process of first selecting one part and then adding another part, and continuing to add further parts until the drawing is complete. This is a bottom-up approach to drawing, contrary to the top-down approach of ordinary drawing practice.

Why, then, are these systems popular? The answer is that these

D - as digitised, no correction

G - correction to grid intersection

L - correct to the nearest existing visible line, circle or arc

M - correct to midpoint of nearest line or arc

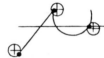
P - correct to existing point: end or intersection of two lines, circles or arcs.

C - correct to centre of circle or arc

O - correct to object origin. (Note this object includes text and has its origin on the bottom of the nut)

B - correct to text block edge.

In certain commands the following are also allowed:

T - construct tangential (or parallel) to line, circle or arc.

N - correct normal to line, circle or arc.

Figure 6.12 *Latching parts*. In drawing a detail, it is often necessary to achieve precise interactions between different line types, as when joining curved and straight lines. The point of intersection can usually be defined in several ways. Some systems offer an extensive range of procedures for defining intersections. Others offer a limited range and rely on the user to combine operations in order to achieve an intended result. In both cases, the user has to think in terms of constructing the drawing using the available computer operations.

Figure by permission of Applied Research of Cambridge.

systems do offer advantages similar to those which support the success of word-processors. Once a drawing has been set up in a computer, it can be edited and transformed for reuse in many variants of the same drawing, and as parts of further drawings (Figure 6.13). This advantage can be applied to drawings of all kinds, including those that do not obviously consist of repetitive elements. All buildings have walls, for example, and all walls have similarity in their depictions, and that similarity can be exploited by drawing systems. This advantage applies even where the walls are otherwise known as being obviously different in each building. An additional benefit, again in common with word-processors, is that when drawings are reused they retain the quality of original

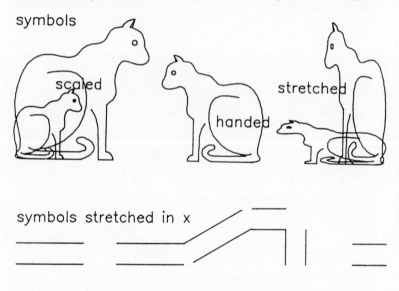

Figure 6.13 *Instances and transformations*. When the same part occurs many times in the same or different drawings, it can be made into a separate drawing and instances of this drawing can then be included as sub-pictures within other drawings. An instance here means a sub-picture, which is invoked by getting the computer to refer to the separate drawing of the part. Each instance can be separately subjected to transformations, such as rotation and stretching, so that they can be different from each other and from their original drawing. The virtue of transformations applied to sub-pictures should be apparent if we consider instances of a wall part rotated and stretched in different locations within a plan drawing of a building.

Figure by permission of DeCAL.

Drawing Systems

drawings. This means that all drawings can exhibit a consistently high standard of presentation.

Further claims for drawing systems tend to be confusing. The claim that systems ensure accuracy through their consistency is subject to a person first deciding that a drawing is correct. Such accuracy is then vulnerable to the effects of the edits and transformations that may be applied to new instances of the drawing. Similarly, the claim that systems employ models that support quantification is dependent on correspondence between discrete picture parts and a user's perception of the depicted things to be quantified. If such correspondence is built into a system, it will make the system specific to particular applications. More problematically, there is the claim that systems employ three-dimensional models to support perspective projections. The essential function of such a model is to collect values that will position points in three-dimensional space, often from two-dimensional drawing input, so that these values can be used by some perspective generating program. As an illustration of what can happen, consider Figure 6.14. Here we have a polyhedron model supporting three-dimensional projections, and we see the effects of the user getting hold of parts and changing them. These might seem perfectly reasonable actions to the user, but the system's restricted model leads to unexpected results. The issue here is not whether we should welcome perspective facilities, but to question the generality of the three-dimensional models that systems employ, as an extension to general purpose drawing systems. All such extensions tend to be specific to certain applications, with ambiguous implications for the distinction between drawing systems and the role of drawings in integrated design systems (Chapter 4).

Drawing space

In computerised drawing systems, drawings appear on display screens. The drawing space consists of a densely packed set of defined points arranged in a square grid pattern which is represented in computer memory. The computer controls which points are displayed on the screen at any one time. Strings of points spaced close together give the semblance of lines. Thus, as seen from inside the computer, the basic drawing primitive is a point and all other drawing objects have to be described in terms of points.

In effect, an architect's understanding of drawings in terms of lines has to be translated into a computer's representation of drawings in terms of points. People's higher level interpretations of drawings as, for example, Euclidean geometry, have to be trans-

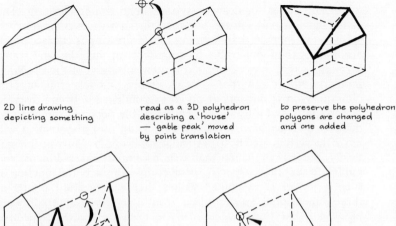
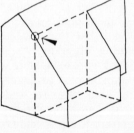

2D line drawing depicting something

read as a 3D polyhedron describing a 'house' — 'gable peak' moved by point translation

to preserve the polyhedron polygons are changed and one added

'house' joined to a 'tall house' — 'ridge' moved to 'eaves' to form a co-planar 'roof', by line translation

requires a location for an intersection to restore vertical plane of 'gable'

Figure 6.14 *Three-dimensional interaction*. A three-dimensional polyhedron model of an object, a house, is depicted in an isometric projection. A designer may know what a polyhedon is, but he may not be aware of what is expected of any interaction with the drawing in order to preserve the model. The designer may have an intention which prompts him or her to move a single intersection. If the system is 'clever' it might preserve the model by introducing a further polygon, which might not be what the designer intended. Possible effects of further interactions are shown in the lower part of the figure. Such situations can be controlled by setting conditions for lines and by getting the system to instruct the user on how to interact, but this would imply particular anticipations on behalf of designers.

lated into the computer's representation of Cartesian co-ordinate point geometry. Transformations which people perceive as being applied to line-based drawing objects have to be translated into transformations which can be applied to sets of points. By now, developments in computing have produced generalised functions for executing many of these translations, so that developers of drawing systems can take them for granted.

Drawing Systems 163

In practice, a computer may have several arrays of points in memory and the number of points in each can be far greater than the number available on physical display devices. Systems are then able to support selective windowing on different parts of computer memory, to bring different graphical and textual images into view (Figure 6.15). Windows can be made to behave virtually like

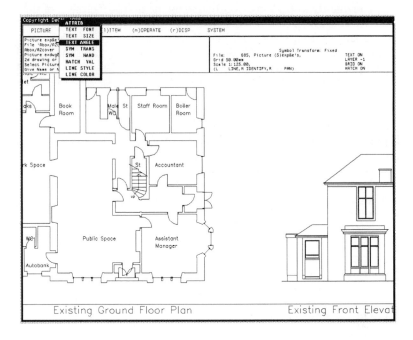

Figure 6.15 *A workstation screen*. A computer workstation is shown with pop-down menus and screen windows used by a drawing system. Menus appear when header items are touched by the screen cursor which, in turn, is controlled by mouse moves. Selections are made by mouse clicks. Menus provide access to computer operations for setting parameters for the drawing space, calling drawing items (line types and sub-pictures), setting parameters for items, invoking edits to items, and for moving around the displayed image. Any selections can be made in any sequence. The system provides screen windows for the menu headers and program status information, and a large window onto the drawing space in computer memory.

Figure by permission of Compass Graphics and DeCAL.

separate terminals which support user interaction with the different programs that constitute a system. Moreover, since the points visible at a screen face are known to the computer, it can translate them to other co-ordinate spaces, and it can subject them to scaling and other transformations. A drawing on the screen can be related to other drawings in the computer's memory and to metric dimensions in the user's world.

The quality of drawings is affected by the point resolution of the devices on which they are displayed. Point resolution refers to the number of points that fill the display space. The higher the resolution, the smaller the distance between points and the more strings of points will appear as continuous lines. Devices with low resolution will display most lines with more obvious steps. This condition applies to all display devices, including printers and plotters. However, since the drawing space is in computer memory and not in its display devices, a drawing can be displayed to the full potential of each device. Thus, for example, the same drawing seen on a display screen can be plotted on paper at a higher resolution, resulting in improved line quality.

The drawing space of a computerised drawing system occurs in computer memory and is seen on computer display devices. This combination of memory and display serves as the equivalent of paper in ordinary drawing practice. Paper, in the latter case, is the medium for both memory and display. The important difference, of course, is that computer memory is controlled by computer operations, and this difference allows us to use a computer to change its memory, to edit and transform displayed drawings. To gain this benefit, we have to accept and work with the computer's point-orientated view of drawings.

Finding your way in drawings
What, if any, are the implications of the computer's point-orientated view of drawings on people's ordinary perceptions of drawing? We can observe that all descriptions (user manuals) of drawing systems present users with the assumption that points are the essential beginning of any drawing, and that pointing is the only way of interacting with drawings. The structure of drawings is presented as hierarchically related points describing lines, collections of lines, and higher level objects such as polygons. All systems require users to select points and then select the drawing operations to occur through or between the points. This emphasis is reinforced by the circumstance that a user's only possible contact with the drawing space is by pointing at it or, more precisely, by moving a

Drawing Systems

'mouse' or other device that controls a cursor's position on the screen face.

A good user interface will mask the system's point-orientated view of drawings from the user, allowing interactions in terms of lines and line intersections. As an alternative to pointing, we might expect people to be allowed to talk about parts of drawings as a way of getting at the parts of a drawing. From our earlier discussions of ordinary drawing practice, we might then expect people to get at any desired selections of lines. However, this ability is precluded by the intended generality of drawing systems.

Here we can note a difference in the way in which word-processors work. If we were to apply the same reasoning for drawing systems to word-processors, we might expect that we could get at words only by pointing at them. Why is this not the case? The reason is that we know a lot more about depictions of words than we do about drawings in general. We can use this knowledge to anticipate properties of depictions of words, and we can build these anticipations into procedures for finding and doing things with words (see further discussion on technology of words in Chapter 8). Thus, we can interact with text stored in a word-processor by means of pointing, by saying what we want, and by combinations of both pointing and saying. We can do this with a dumb word-processor that does not know what the words mean.

For drawing systems, we do not have equivalent knowledge about drawings that would tell us how drawings decompose into parts and how we might recognise the parts. Let us be specific about what we do not know. To achieve the equivalent functionality of word-processors, a drawing system should allow us to draw a shape, any shape, and then ask for any other instances of the same shape that might be found elsewhere in the drawing or in any other drawings. Instances of the shape we are looking for may exist elsewhere in the drawing and in the other drawings, but they may not have been entered as separately defined parts of those drawings. The system will then not be able to find what we are looking for.[1]

For the time being, we can recall drawing parts only if they happen to be general primitives of the system, such as lines, circles, and boxes, or if they have previously been entered to the system as separately named drawings. Otherwise, we have to be content with pointing, to redefine to the computer all the intersections that describe the part we want. In this respect, we can conclude that drawing systems are less powerful than word-processors.

Instructing the computer

The criteria for a good drawing system are that it must be capable of producing any drawings you want to the complexity and detail you need, reliably and efficiently. To meet these criteria, a system must be prepared to receive *instructions* from you. The system developer must anticipate what those instructions will be and build that anticipation into the program. Such anticipation is unavoidably prone to being wrong. The aim in designing a system is, therefore, to minimise anticipation and maximise exploitation of the system's procedures. These are general considerations which affect the view of a system as seen by the user, and they affect the actions required from the user to run the system.

Generally, to produce a drawing, the system requires instruction on where to place a *line* and where it stops. These instructions have to be given for every next line. By repeating this procedure many times, it is possible to achieve elaborate drawings incorporating straight and curved lines of various thicknesses (and colours), including variously dotted and dashed lines, hatched areas, and text annotations.

The repetitive procedure for entering each line (or hatched area, or text item) is inherently onerous. System designers seek to reduce this burden in two ways. First, the *input procedure* can be simplified by ensuring that maximum information is received by the computer from minimum action by the user. This can be done, for example, by setting line types and styles for sequences of lines, and by positioning, orienting, and delimiting successive lines to successive screen locations indicated by the user.

Secondly, and more effectively, any collection of lines forming *partial drawings* can be identified and stored for reuse in the same or any other drawing (Figure 6.13). This allows, for example, layered or nested drawings within drawings any number of times. When recalled, partial drawings can be subjected to transformations so that repeat instances are no longer identical. The procedure for defining, storing, and recalling partial drawings (sub-pictures, picture components, etc.) can be simplified so that the effort is similar to entering a single line, and the procedure should be operable at any time during the normal course of drawing.

Systems vary in the amount of help they provide for *positioning* drawing parts in relation to other parts, like rotating one part with respect to another, or fitting curved lines to straight lines (Figure 6.12). The potential range of procedures is very large, especially if alternative procedures for achieving the same end result are included. Systems which respond to separate instructions for each procedure are, initially, easy to use but eventually they require users to know a large number of instructions for a comprehensive set of outcomes. Furthermore, users tend to lack access to the full potential of the system. The alternative strategy is to provide a relatively small set of basic procedures and users then employ these in different combinations to achieve required end results. These systems require more understanding from their users, but they are easier to operate by experienced users. These users have more control in making the system produce their own intended results.

Dimensional precision in placing drawing parts is generally achieved by latching them onto intersections of previously positioned lines (Figures 6.11 and 6.12). Systems offer various procedures for setting up regular patterns of construction lines as *grids* with known values for grid intervals. Grid value and orientation should be variable at any time during the course of drawing to provide any required dimension for any drawing part. The grid on a display screen is usually controllable to set very small or very large values.

Changing your mind
Given a drawing that already exists in a computer, and given that the drawing depicts some state of a design object during the process of being designed, it is certain that the user of the drawing system will want to *change* the drawing. Moreover, given that the user is designing, it is probable that the user will want to make unpredicted changes to any part of the drawing, including parts that might not previously have been identified as distinct parts.

To execute any change, the designer has to indicate the part in a manner that will define the boundaries to the part which is to be changed, to the system. As explained earlier, the designer can call for a part by name only if it is in a previously named part. The name must refer to some existing part of the logical *storage structure* by which the system holds the drawing. The structure will usually include system-defined names for line types and styles, and user-declared names for assemblies of lines into sub-pictures, and for assemblies of sub-pictures in larger compositions. These logical parts are hierarchically related in the storage structure and they determine access paths to parts of drawings. Instances of these logical parts can occur repeatedly as different parts of a drawing and in different drawings. Usually a system can recognise whole instances of parts in context.

The designer, as a user of a drawing system, is now faced with two problems. First, a displayed drawing will not normally reveal the structure by which it is know to the system. The designer might not remember or might not know the structure that will have resulted from the particular instructions used to enter the drawing into the system. Most systems acknowledge this difficulty by allowing the user to call for *visual clues* to a current structure. These clues may appear as origin points for all sub-pictures, or all instances of a sub-picture, within the whole or any part of a drawing. The origins may also be displayed for nested levels of sub-pictures. A system can be asked to highlight all the lines that are parts of a single sub-picture. However, few systems are able to reveal current attachment conditions at intersections of lines, affecting what might happen if only some are moved.

Secondly, and more seriously, the collection of lines that forms a part which the user wishes to change may not coincide with any part known to the system. The part might be clearly visible to the user, but the system will not be able to recognise it. The user might be employing a *decomposition* that refers to parts of line lengths, angle values that are inferable but were not previously specified, and attachment conditions that do not correspond to a single and whole previously defined sub-picture. In all such cases, the user has to cross between levels in the system's storage structure, to uncover the lines that he or she wants, and then reassemble the selected lines into a new sub-picture. Such operations can be computationally expensive and very demanding on the user's knowledge of the system.

This difficulty is serious because any requested change, such as a line edit or a shape transformation, can be executed by the system only on discrete parts as known within a current structure. Changes can be executed only uniformly on whole line entities, sub-pictures, or larger compositions. Some systems try to deal with this difficulty by forcing decompositions that are considered to correspond to parts of design objects, expecting a designer to compose a design from a library of predefined sub-pictures. Such a system is dependent on correct anticipations of design objects, which then have to be widely acknowledged by all designers who are potential users of the system.

This review of drawing systems provides an indication of the kind of effort required to use them. Drawing systems have not yet achieved the kind of generality that has been essential to the success of word-processors. Some might argue that drawing systems will never achieve this generality. We might have to accept that drawings are inseparable from domain knowledge, as a consequence of their role in depicting such knowledge. If this is the case

Drawing Systems

then the problems presented by drawing systems have interesting implications for our general ambitions for computer technology. How can we expect computers to be generally useful for expressing and operating on any expressions of knowledge from any people? This question will be considered further in Chapters 7 and 8.

Choosing a system
Despite the reservations set out in this chapter, drawing systems do offer designers one of the least threatening routes for entry into computer technology. The following paragraphs will note some practical considerations affecting the choice of any particular system. Any prospective first-time user is immediately faced with a general difficulty. You will not already have experience of using computerised drawing systems, so you will not have a precise anticipation of what you can expect from a system, and you will not be able to make a calculated comparison between alternative systems. Vendors, either unintentionally (through not knowing what your needs are) or intentionally (they have to sell their systems) are likely to be misleading.

In the absence of experience, choice has to be guided by three primary and general considerations that apply to any use of any system. First, design practices change. This means that you need a system which is least likely to inhibit change, leading to a preference for the most advanced but tested technology that you can afford. Secondly, in time your choice will be wrong. This means that you must limit your investment to a level that you can afford to replace. Thirdly, the major part of your investment will be in learning how to exploit your computer environment, which is an investment that you will want to keep.

These three factors are somewhat contradictory. The last is the most critical. In the case of architects using drawing systems, more than a year can be invested in learning how to use and exploit a computer, and in making consequent adjustments to existing practices, before a system becomes an integral part of the user's design practices. Such investment is necessary before the system can be trusted to produce drawings that will represent your design knowledge to the world outside your office.

Limiting investment in your acquisition of a system is less critical, since this is likely to cost less than your investment in learning to exploit the system. If you think that your drawing needs are modest, then a system running on an IBM PC environment might be sufficient, and the alternative Apple Macintosh environment may be more appealing (see comment under progress, in Chapter 8).

However, if you intend to use your computer to produce full-scale working drawings for buildings, in place of all drawings that you would otherwise produce by hand, then you will probably require a more powerful computer environment.

Once your investment is made, you will want to preserve it against incompatible new developments in computer technology. But your evolving design practices will make new and unforeseen demands on your system. It is likely that you will want your system to be reprogrammed, to take advantage of new possibilities for relating computer-produced drawings to your own design practices. It is this prospect which favours the choice of the most advanced technology that you can afford. An example is offered by powerful Sun/Unix workstations, which use a computer operating environment that is being adopted by many computer manufacturers and that is used by many researchers engaged in fundamental advances in computer technology.

Notice that I have not stressed the need to choose a particular software system, a drawing program. You should not expect a close fit between a particular system and your own requirements. It is more important to select a computer environment which will allow you to switch between different drawing systems, or allow a particular system to be tailored to your own changing needs.

SUMMARY

This chapter has focused on that branch of CAD which is currently popular and offers practical applications: drawing systems. We have drawn parallels with word-processors and have explored the virtues of dumb systems. Advantage is gained by treating drawings as things that can be stored in computers, recalled, subjected to edits and transformations, and reused. Their generality and, therefore, their potential for widespread application are dependent on drawing systems not knowing what drawings depict. This absence of knowledge avoids conflict with the various states of knowledge within different users.

We discussed the symbolic nature of words and the analogic nature of drawings, and considered how they are employed in ordinary drawing practice. Drawings are a pictorial form of expression, used to express spatial properties and relationships of depicted things. This makes drawings vitally important to designers. The act of drawing needs an environment in which people can depict their own thoughts and evoke corresponding thoughts in other people. In considering computerised drawing systems, differences in our general understanding of words and drawings were related to the possible functions of these systems. Limits to the regularities which occur in drawing practice should lead us to expect limits to the range of tasks performed by these systems. Increasing their prescriptive functionality, by encoding higher level knowledge about depicted objects, would have the inevitable effect of reducing their acceptability among people in the same or different fields of application.

The use of computerised drawing systems differs from ordinary drawing practice: users have to think of drawing as a deliberate act that is separate from thinking about depicted things, and drawing procedures have to be executed in a logically structured environment. Nevertheless, such drawing systems exist, they are useful, and they are continually being developed to make them more acceptable to designers. This chapter concluded with some practical considerations, stressing the advantage of choosing a computer environment that offers a path to new advances in software and hardware technology, and one that can be tailored to the user's own changing design practices. The next chapter will consider the possibility of future representation schemes which will allow users to re-establish the link between drawings and depicted things, in computers.

Notes to Chapter 6
1. Perhaps the most promising area of research engaged on the problem of getting a system to find its way to arbitrarily declared parts of drawings, is the field of shape grammars described by Stiny (1980).

CHAPTER SEVEN

DESIGNING IN WORDS AND PICTURES

If it can take anything I say, but only if expressed in terms it already knows, then how can it get it to know what I know?: here we will consider how design descriptions, expressions and concepts might be represented in computers, illustrated by a worked example.

The previous three chapters have described experience of CAD applications: integrated design systems, function-orientated systems, and drawing systems. Given the general intention that computers should serve and support the work of designers, as a group of people who in turn serve other people, the previous experience exposes fundamental questions.

Integrated design systems depend on a close correspondence between formalised models of design practices within computers and individual designers' perceptions of their own practices in a world in which designers have to be responsive to unforeseen demands from other people. Is it possible to conceive of computer systems which can evolve with designers' changing practices?

Function-orientated systems are focused on discrete aspects of design practice, but the tasks they perform have to be recognised as being useful to designers, which means that results have to fit in with designers' intentions for whole design objects. Do task-specific systems have to know their users, and can they do so?

Drawing systems offer a means for producing descriptions of design objects, but these systems do not know what the objects are outside the drawings, and the operations they offer condition the range of objects that can be drawn. Is it possible to conceive of systems which will re-establish the link between drawing and thinking about design objects?

Earlier chapters considered more generally the role of language in externalising knowledge, and what we can know about

Designing in Words and Pictures

knowledge. We concluded that computers essentially are mechanistic devices for processing symbols, without knowing what they are doing, at least not in the human sense of knowing. Despite this restricted view, computers can be made to do impressive things, and their ability to process symbols can be regarded as an extension of human literacy. How, then, can computers be made to serve designers?

In this chapter I will draw together these two themes, linking experience of practical applications to a more theoretical consideration of computers. To do so, I will restate the earlier general arguments in a manner directly relevant to the task of expressing design knowledge, describing design objects. I will try to establish a position somewhere on a spectrum of beliefs between two extreme and implausible positions. These extremes are, on the one hand, the belief that anything that might occur within people must exist in a manner which we cannot explain and cannot represent logically outside people. On the other hand, there is the belief that the totality of all that can occur within people potentially can be represented wholly in computational logic. Both of these positions seem implausible, but they are useful reference points for any declaration of a plausible ambition.

My further argument will rest on the apparent fact that commonly people do use formal systems, as in the use of written language, and that this is part of being human. It is also apparent that this usage is not exclusive; it refers not only to logically substantiated knowledge but also to other unexplained human sensitivities — so we are able to talk about ethics and aesthetics. People use formal systems to evoke these human sensitivities. Usage extends beyond the formal definitions of systems and usage has the effect of redefining these systems. This interplay between formal systems and other human sensitivities is an essential focus of CAD.

PRESCRIPTIVENESS

All the questions posed by experience of CAD point to a single issue: the prescriptiveness of formal systems. Prescriptiveness refers to a distinction between knowledge as part of people and representations of knowledge outside people, as outlined in Chapter 2. The issue of prescriptiveness refers to limitations on our ability to establish correspondence between:

(*a*) people's individuality, which conditions what we can know about things;

(*b*) formalised concepts represented in computer systems.

Concepts are considered here as formalised expressions of people's individuality, or people's notions, as in the form of written words. Formalised concepts serve as representations of people's notions. As representations, they are not the same as the things they represent. When we use a computer system to represent concepts, and operate on those further representations, we have to be satisfied that the system behaves as though it were using our notions. Ideally, we want to be able to define and redefine a system's concepts to be in tune with our notions.

Prescriptiveness of computer systems refers to the encapsulation of partial knowledge in computers in a manner not modifiable by their users. Only the consequences of such encapsulations are visible to users. Furthermore, prescriptiveness includes system-defined interpretations for any behaviour of people (including 'user error' states) when they use computer systems. Prescriptiveness sets boundaries to the things people can do with computers. These conditions hold even when we are unable to determine those boundaries or predict the behaviour of users.

Computers and other systems
With these observations in mind, we should note that prescriptiveness is not unique to computers. It is a property of all formal systems exercised by people. It occurs in our use of familiar expressive environments to support language expressions in the form of text and in drawings. If we were to get rid of all prescriptiveness, our language expressions would become uninterpretable and meaningless.

The next significant point to note is that prescriptive systems generally are exercised within people. These systems can come into existence without premeditation, they can be used unknowingly, and they can be evolved through usage. In most cases, people's use of a system is not bounded by its prescriptiveness. People are able to make references outside a formal system, which can have the effect of redefining its prescriptiveness. Here, computers do present a new problem. Any computer that performs a useful task cannot follow people's references to experience outside the prescriptions which define the system. Normally, in practical cases, people as users of the computer are not able to redefine its prescriptiveness; to redefine the machine.

This limitation on computer systems refers to their inability to share the same direct world experience of people. This limitation could be overcome only if computers were to become so like people that they would *be* people — a profoundly presumptuous ambition. The problem of prescriptiveness within computers is likely to become more severe the more computers are made to perform domain-specific tasks, meaning those tasks which people are expected to recognise as being equivalent to human tasks in application domains. The problem is likely to become more severe the more computers are believed to match human intelligence.

It is this line of argument, supported by experience of CAD, which leads to my advocacy of people's intelligent use of dumb machines. This chapter will explore the implications of this ambition.

Descriptions

Designers produce descriptions of design objects. These descriptions have to account for the production, performance, and maintenance characteristics of designed objects. These objects can be buildings, or they can be any other concrete or abstract thing. In all cases, descriptions have to exhibit design knowledge about the behaviour of design objects, and this knowledge has to be responsive to the diverse interests of other people for whom the objects are designed.

In seeking a general characterisation of descriptions, we should be guided by three primary concerns. First, the broad intention is to discover how designers, in general, can use computers. Secondly, designers tend to express their knowledge in the form of drawings, which prompts a focus on systems for producing and modifying drawings. Thirdly, experience of design practices shows that human actions are not determined by externalisable logic, which prompts a focus on interactions between design descriptions and logical operations within computers. Out of these concerns, we want a definition for descriptions which can serve to define a description system.

In working towards a definition, we can say that descriptions occur temporally in some current context, without any external and absolute authority and, therefore, without any notion of being objectively correct. Descriptions represent current views of the world as known by the people (or computers?) who produce them. A description might refer to knowledge of situations in the past or to intentions for the future, but in these cases the description itself describes current knowledge or intentions. This temporal property of descriptions carries the important implication that descriptions should be expected to change through time. A description system is one that can accommodate unforeseen changes.

The temporal property of descriptions also carries the implication that descriptions remain the immediate responsibility of the people who produce them. Descriptions are used by people to manifest their responses to each other. To serve as a means of communication, descriptions must have a physical presence outside people.

Descriptions themselves have to be composed of objects and relations that constitute a description environment. These objects and relations have to be present in a form which can be subjected to actions executable by recipients, by people, and by computers. A description system, then, has to consist of prescriptions for objects and actions that apply to these (and only these) constituents of a description environment.

The prescriptiveness of a description system cannot account for anything that might be known outside the description environment, outside the system. A description system cannot carry responsibility for anything a description might describe. The goal, then, is a set of prescriptive functionalities within computers at a low enough level to serve as an environment in which people can exercise and exhibit intelligence. The aim is a computer system which can be made to reflect unexplained human intelligence.

Here we have outlined a definition for descriptions which corresponds to expressions as discussed in Chapter 2. Descriptions are expressions. However, the emphasis intended for descriptions is on the formation of expressions, the making of expressive objects independently of interpretations. From the earlier discussion on language, such a dissociation of expressions from interpretations must result in meaningless expressions. This is precisely the intention for description systems. Again, from earlier discussion, a computer system does not have interpretations or meanings for any expressions it is given, in the sense that people know meanings. The system has only received meanings between expressions and, from its point of view, the expressions remain meaningless.[1] The expressions ought to be, and need only be, meaningful within the people who use the system. A description system, then, is one which people can use to produce and modify their own expressions.

This distinction between constructing and interpreting descriptions can be difficult to maintain. The use of a computer to construct a description can invoke machine behaviour which looks like it understands what it is doing. However, the emphasis on constructing descriptions remains valid since this is vital to the ability of people to describe changes in human understanding, either within or outside computers.

Objects
This printed page offers familiar examples of objects composed into

descriptions. The text serves as descriptions of the state of knowledge in my mind. The task of constructing descriptions utilises known objects, relations, and rules to form character strings. Changes applied to these character strings reflect changes in my knowledge. My ability to produce and modify such descriptions is dependent on my knowing about the structure of the description environment, about the constituents of this environment. This knowledge needs to be (relatively) independent of my further knowledge about other things I might be describing. The rules which apply to objects and relations within this description environment are not the same as the rules which govern interpretations of descriptions into domain knowledge outside this environment.

Now we have a position which sees descriptions as being composed of objects that exist outside people, as compositions produced by people. The environments in which descriptions are constructed also exist outside people. Examples of these description environments occur as bits and structured arrangements of bits in computers, as characters and character strings in text, as lines and assemblies of lines in drawings, or, indeed, as pieces of plasticine sculpted into solid objects.

My purpose in presenting this position is to emphasise a distinction between making things (such as descriptions) and thinking about things (as when we interpret descriptions). In the example of text we have the production of written words as distinct from thinking about what to write or read. Text is produced within an environment of writing, defined in terms of characters and character strings. In using that environment we employ a system. We find that system useful. We also find machine implementations of that system useful, despite their having no knowledge about our thoughts. In extending our use of computers, we should expect similar usefulness from different description systems supported by computers, especially if they include the construction of drawings.

Functions

In using a computer to support a description environment, we expect it to employ computational functions in the course of producing or modifying descriptions. As noted before, the functionality of a computer has to be defined in terms of some externalised logic. Precise definitions of functions, such as those that occur in formal logic and mathematics, refer to mechanisms within certain representation environments. These functions support representations of certain kinds of knowledge, but not all human knowledge. They offer mechanisms for establishing truth conditions, but these truths commonly are not the same as humanly perceived truths. Knowledge which conditions human actions is not limited to knowledge which can be verified by computational functions.

A description system is one which employs mechanistic functions solely for the purpose of producing and modifying descriptions, including the execution of mappings from one description, in one form of expression, to another. These functions have to be defined so that they remain accessible to people who use them, and are controllable by people when they use them. In using such functions, we should be able to use computers to produce descriptions of anything.

Expressions

Expressions, as discussed under language, are conditioned by interpretive intent. Thus, expressions are part of language as used between communicating parties, between like-minded people. Expressions evoke interpretations within people which involve actions within individuals. In this sense, interpretations are more than mappings between expressions. They might invoke actions that effect new mappings.

Expressions, seen in this way, present problems when we intend to subject them to computational treatment. A description system in a computer has to employ a distinction between those expressions which have interpretations for people only and other expressions which can also have system-defined mappings for machines. This distinction has to remain visible in any descriptions realised by a machine system. In CAD, our aim must be to enable designers to exploit system-defined mappings to generate expressions which can pass between people.

Concepts

Concepts, as in the example of words and word constructs, are expressions which are used as though they have meanings which are the same for different people. Concepts are formalised expressions of people's individuality, and they are important to our ability to construct our shared inter-subjective reality. We use words as concepts that are manifest in language.

To serve as concepts, words and word constructs have to be realised in accordance with some formal structure. Expressions have to have decompositions into parts. Parts have to be known in terms of the roles they can play in compositions, as in parts of speech. We can then use parts to compose expressions which

Designing in Words and Pictures

represent the individuality of persons, or meanings within minds. We can also look at parts and their relations within expressions so as to interpret them in meanings in minds. This position accords with the people-centred view of knowledge and language presented in Chapter 2. As noted before, we cannot know if and when meanings actually are the same within any two people; that is why we go on talking and why knowledge has to evolve with the evolution of people.

To be effective, the structure of expressions that serve as realisations of concepts has to be fairly stable. We have to formulate expressions in fairly consistent ways, in order that we can use each other's expressions. In as much as meanings can exist outside people, we can consider the meanings of expressions to be their structure, defined in terms of decompositions and the roles attributed to their parts. For natural language systems, these parts usually have to be known as entries in a lexicon so that they can be subjected to rules of grammar. The success of expressions in serving as representations of meanings within people depends on the extent to which the parties to a discourse share the same lexicon and grammar.

A usual strategy for natural language systems is to devise a model of the users' world, represented in a system's knowledge base, and define roles for lexical items in terms of the model. The broad idea is that if a model can be made to correspond to the users' view of the world, then the system's use of lexical items, its application of grammar rules to those items will result in linguistic behaviour that can correspond with the users' use of concepts. The fundamental and speculative ambition is that people and computers, when they are engaged in a discourse, should share the same use of words.

In CAD our strategy has to be a bit different. Our ambition is to enable people to construct their own models within a system, and to use their models to produce expressions that can be passed between people. Responsibility for formulating a model and interpreting expressions produced from the model, with respect to any world in which a model might apply, rests with people who use the modelling system. In such situations we have to focus on concepts that condition the production of expressions, without the help of discourse domains already represented in computers. We have to focus on concepts that can be used to describe anything.

Using a computer as a modelling environment, we have to establish correspondence between concepts as manifest in surface expressions which pass between the computer and a person and symbolic representations of these concepts which can be passed down into the computer's knowledge base. This general position is shown in Figures 7.1 and 7.2. These symbolic representations have

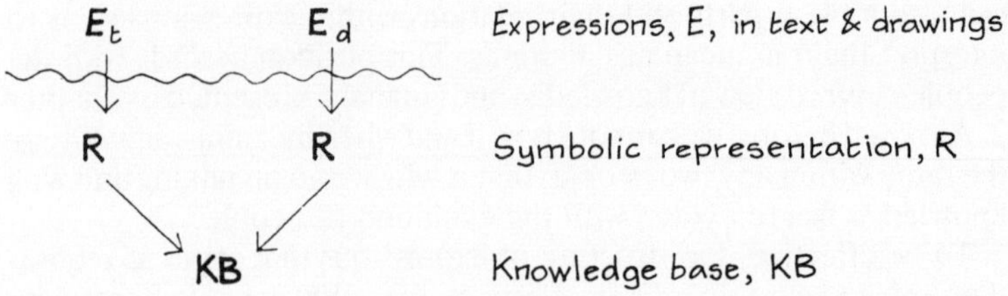

Computer Discipline and Design Practice 180

E_t E_d Expressions, E, in text & drawings

↓   ↓

R   R   Symbolic representation, R

↘ ↙

KB   Knowledge base, KB

Figure 7.1 *Surface expressions*. Concepts are realised in surface expressions, as in the form of text or drawings, which reflect states of knowledge within computers and people. In the case of computers, surface expressions have to be represented symbolically, and decompositions have to invoke machine functions for establishing logical relationships in a system's knowledge base. Responsive behaviour from the system might look as though it is using human concepts.

to correspond to decompositions of surface expressions and they must be able to invoke machine functionalities which can match the roles associated with parts of expressions. Here we still need lexical items and grammar rules, but both are targeted more at the production of expressions and are domain context free. Moreover, symbolic representations ought to be full 'copies' of surface expressions so that anything a person might do with an expression, to modify it, can be reflected in the computer's knowledge base. This ambition is, of course, not attainable — a person using a system has to operate within the bounds of the representation environment defining the system. However, this ambition is useful if it makes us strive for systems that can accommodate maximum freedom of expression with minimum interpretive responsibility. Thus, we arrive at a strategy for CAD as indicated in Figure 2.2.

Figure 2.2, in Chapter 2, includes expert modules as users of graphical concepts which support expressions which pass between a computer and its users — designers. Such expert modules can be viewed as descriptions which have been formulated by people, which have become accepted as established overt knowledge, as in the case of three-dimensional geometry models. In principle, there is no firm distinction between expert modules and anything else that might be described in a computer. This position will be illustrated later by the worked example. Meanwhile, we will allow the possibility that people might accept machine behaviour which is conditioned by expert modules and in these situations the modules can be regarded as users of graphical concepts to make behaviour evident to designers.

Designing in Words and Pictures

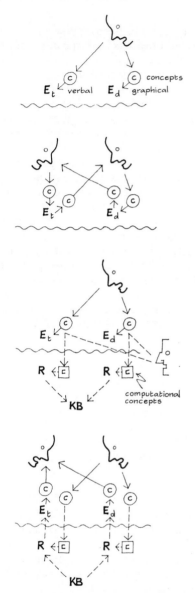

Figure 7.2 *Use of concepts*. The transition from human concepts to formal representations is illustrated. Designers require formalised concepts which they can use to support their own interactions with other people.

Graphical Concepts

Our focus on designers results in an emphasis on drawing as a mode of expression. Design drawings have meanings in terms of spatial properties and relations of depicted objects. These further objects are also known in terms of their non-spatial contexts. These contexts include the circumstances in which design objects have to be produced and in which they have to function. Typically, drawings have to refer to knowledge about production and performance requirements that is not apparent within the drawings. Designers use drawings in conjunction with non-graphical expressions — words and numbers — and designers are able to effect mappings between these modes of expression.

Graphical concepts refer to the ways in which people associate meanings with drawings, without necessarily translating drawings or parts of drawings into words. These meanings refer to spatial properties and relations visible in drawings. By labelling these meanings with words, it does not follow that all values for properties and relations can be expressed independently of a drawing. Typically, values will remain in a drawing and will not be carried across into some other formal interpretation of the drawing. Finding a value might then entail looking at the same drawing and reconstructing a different interpretation.

People are able to reason with values they see in a drawing (... this part of this picture ... here, has this relation with respect to that ... so that it [the depicted thing] will match that [a non-graphical context] ...). This ability to work directly with values of properties and relations that are manifest only in drawings (or in other representation environments in which the spatial arrangement of primitive elements is significant) can be called spatial reasoning.

If we intend to support this ability by using computers, we need to be able to describe graphical concepts (the meanings of arrangements of line properties and relations, in terms of spatial states like on, next to, near ...) within computers, to provide them with a basis for manipulating graphical expressions. These concepts need to be developed so that drawings can be made to interact with other knowledge representations. The results of spatial manipulations should be allowed to alter the contents of a non-graphical representation and, conversely, changes to non-graphical knowledge should propagate changes to a graphical representation, to a surface expression in the form of a drawing.

To support this two-way correspondence between graphical and non-graphical representations, graphical concepts need to be related to other concepts which apply to text and numeric expressions, which are used to refer to domain knowledge. Graphic concepts have to be useful for referring to spatial properties of those domain objects and operations relevant to the production, maintenance, and performance of designed objects. Formalisation of such graphical concepts, integrated with verbal concepts, is a major ambition.

Graphical productions
I have said that abstractions from expressions are conditioned by knowledge about expressive environments. These environments consist of definitions which effect the production of expressions. Production refers to the making of expressions, realised as expressive objects, and not to the act of deciding what particular expressions should be. Definitions of production environments determine what expressions can be, and influence the possible range of interpretations for those expressions. This position holds for both text and drawings, as illustrated in Figure 7.3.

Text, as depictions of words, employs character strings in the form of one-dimensional line graphics. For character strings to be interpretable as words, they must be present as parts of single linear strings. By satisfying this condition, it becomes possible to devise logical abstractions that capture much of the significance conveyed by text. Of course, these abstractions require lexical, sentential, and intersentential syntax, but all these possibilities rest on the linearity of text expressions.

Drawings as depictions of spatial properties and relations of other concrete or abstract objects employs multi-directional linearities in the form of two-dimensional line graphics. Drawings can only be two-dimensional objects (even when drawn on curved surfaces). Interpretations for drawings have to refer to the structure of possible arrangements of lines: to knowledge about how lines can occur in a depiction, and to the ways in which line assemblies can be decomposed. This condition is obscured by people's familiarity with particular drawing practices — they take structure for granted, which presents problems when they come across unexpected structures in particular computerised drawing systems. We need to aim for explicit definitions which can constitute a structure for drawings, which can be accepted as general to different drawing practices.

This condition on line drawings also applies to all two-dimensional graphics, including depictions which consist of brush strokes and area rendering. Depictions remain two-dimensional objects.

1-D LINE GRAPHICS

eg

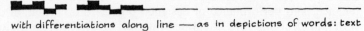

with differentiations along line — as in depictions of words: text

a graphical (text) production machine
    allows addition and deletion
    of discrete graphical entities
    along a single linear path

2-D LINE GRAPHICS

eg

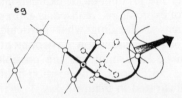

with differentiations over 2-D surface
    exhibited by angle, length, curvature... values: drawing

a graphical (drawing) production machine
    allows edits that propagate changes
    to non-discrete graphical entities
    in any direction across a single plane

3-D MODELLING

eg

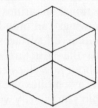

this depicts a cube
    if it can be made to correspond to
    a model:
    a solid that is completely bounded
    by six flat sides and each side is
    an equal sized square

the model exists outside the graphics
    recognition of an instance of object accommodated by the model
    relies on interpretations from arrangements of lines
    to the entities by which the model is known

Figure 7.3 *Two dimensional graphics and three-dimensional models*.

Designing in Words and Pictures

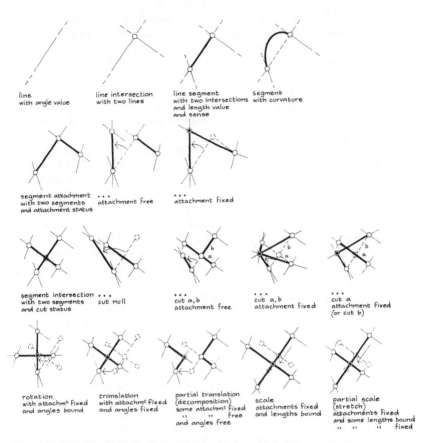

Figure 7.4 *An 'ABC' of line graphics.* This illustrates an understanding of objects which form graphical expressions, based on line entities with angle values and segments with length values, plus intersections and attachments of these entities. A graphical production machine, defined on such knowledge of the structure of drawings, can be instructed to produce depictions of other concrete or abstract objects. This ability relies on satisfactory references between representations of semantic domains and the drawing machine's internal syntactic structure.

Interpretations from any drawing entail reading the structure exhibited in the drawing and establishing correspondence with other things that are not the drawing. Thus, we can think of interpretations to expert modules as shown in Figure 2.2, and a three-dimensional modeller is a common example of such a module. A drawing cannot be a three-dimensional model, but it might be a two-dimensional depiction of a three-dimensional model. Our aim, then, is to avoid structures of drawings which are determined exclusively by the needs of one particular expert module, and to devise a general structure which can serve different expert modules, including as yet undefined modules.

Structure of drawings
Figure 7.4 presents a tentative general structure for drawing objects. This structure accepts lines as the basic entity of drawings. Instances of lines are composed into arrangements which include intersections and kinds of attachment between lines. Line segments can have properties such as curvature, style, and colour and, under appropriate attachment conditions, lines can delimit areas of rendering. All these features of the structure of drawings can be defined independently of any interpretive intentions for depictions: they can constitute a drawing production machine, and they can be used selectively to support different interpretations.

To achieve this generality, it is necessary to target definitions at the lowest visible entities that contribute to the presence of drawings, that contribute to knowledge about marks on a two-dimensional drawing surface. This is necessary in order to be able to decompose any drawings into any parts, including multiple and contradictory decompositions, in the course of establishing intended interpretations. This does not mean that every use of drawings has to employ those low-level entities — an application might be satisfied by higher level primitives (or sub-pictures) which have been separately defined in terms of the low-level entities. However, the possibility of access to low-level entities becomes necessary if an application's view of drawings changes, or if the same drawings are to be viewed from different applications.

Definitions set out in Figure 7.4 see lines as basic entities that are differentiated by angle value. Lines with different angle values produce line intersections. Pairs of intersections along the same line can be used to delimit segments. Segments are further differentiated by length value (and sense indicating direction from intersection to intersection), and segments serve as spacers for intersecting lines.

This basic position differs from most CAD systems that start with an xy co-ordinate point space. It allows for x and y axes, as lines with calibrations, but it is not biased towards those lines or consequent paraxial transformations.

Segments can be joined together at line intersections, under various attachment conditions which affect the outcomes of subse-

Designing in Words and Pictures

quent edits. Segments can intersect each other with different cut conditions also affecting the outcomes of subsequent edits. Cut conditions might include latching of cut ends to the line of the cutting segment (not shown in the figure).

Assemblies of segments, under controlled conditions of attachment and with selected angle and length properties bound (or parameterised), can be subjected to transformations to effect rotation, translation, and scaling. These transformations can be applied partially to effect decompositions, localised distortions, and stretching. These effects can be executed uniformly on any assemblies of lines presented in any orientation, as indicated in Figure 7.5.

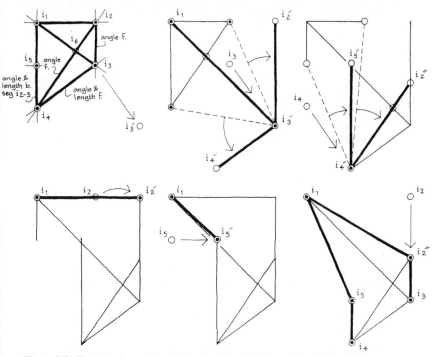

Figure 7.5 *Changing shape.* A shape may be known to a drawing machine as an arbitrary chain of segments with some of their properties fixed and others parameterised (top left). A user might want to stretch the shape by pulling at one corner (i_3 to i_3'). The machine executes the change by a sequence of transformations to arrive at a result (bottom right). It first translates intersection 3 to 3', moving the attached ends of segments to the new intersection (line angles are rotated and segment lengths changed). Any fixed or parameterised values of the changed segments are then restored, detacing their far ends. Loose ends become the targets for further intersection translations (i_4 to i_4', i_2 to i_2' etc), until all attachments of the initial shape are restored in the changed shape.

These general definitions can be represented symbolically in a computer so that it can function as a drawing production machine.[2] The symbolic representation can be used to store and regenerate instances of drawings, in a manner which will allow different and not necessarily related threads or sequences of expressions to be extracted from instances (for purposes of logical resolution), corresponding to parts of drawings. The aim here is that the representation should serve as a passive and 'complete' reflection of drawings, allowing actions initiated by other parts of the system or by a user to set conditions for partial views of drawings necessary for intended outcomes. The drawing machine can then execute any decomposition on any drawing, provided that the intended action is definable in terms of this structure for drawings. Taking the classic example of a pair of crossing lines, shown in the lower part of Figure 7.4 as intersecting segments, the drawing machine can be asked to see this as two, three, or four segments, as pairs of segments forming corners or tees, as groups of three segments forming tees, or as a single cross.

Completeness of a symbolic reflection, or copy, of a drawing is not a well-defined goal, since it depends on the range of possibilities that any person might see in any drawing. We cannot know of a natural and absolute decomposition of drawings. Nevertheless, this goal is important to the definition of a drawing machine which can be used by different expert modules (Figure 2.2) and by different people, for different applications. It is important to the generality of the drawing machine.[3]

Interpretations and drawings
Given a sufficiently general structure of drawings, it should then be possible for a user of the CAD system or the various expert modules of the system to communicate intelligently with the drawing production part of the system. The user, or any other system component, using his, her, or its own expressions should be able to specify changes or transformations to spatial properties and relations of a design object to the drawing machine, and expect it to execute corresponding operations on drawing objects. The drawing machine must have the ability to operate and the user's expressions then serve to invoke drawing operations. The claim for intelligence made here means simply that different components of a system have mappings for each other's expressions which result in acceptable responses.

The usefulness of graphics then depends on links being established between spatial properties and relations which can be depicted in drawings and knowledge about the properties and behaviour of depicted things.[4] For formal systems, these links between depictions and depicted things require something analogous to a text lexicon. Graphical concepts have to serve in a manner similar to the functioning of any lexicon which supports interpretations of text expressions. Graphic items can be regarded like word items, differentiated by multi-directional arrangements of lines rather than the differences in characters which occur in uni-directional character strings. When we have such graphical concepts, we can then have graphical expressions which can be subjected to linguistic treatment. It is only then that graphical expressions can be used by expert modules, as shown earlier in Figure 2.2.

Graphical concepts have to be expressed in the form of drawing objects which conform to the general structure of drawings and operations which the drawing machine can perform on drawings. These expressions have to be related to other expressions in a form 'familiar' to graphics users: other parts of the system; and people who use the system. These relationships can be defined at a low and general level in a manner which is fairly transparent to users, provided that somewhere the system has mappings from high-level and application-specific terms to the low-level terms. These concepts might be expected to include:

(a) items as structured spatial arrangements of parts of drawings (assemblies of line segments with their properties of angle, length, and attachment);

(b) properties which define degrees of dimensional freedom allowed for these arrangements (free, fixed, or parameterised, which condition the effects of spatial transformations);

(c) relationships which are permissible or required with respect to other items (conditioning spatial compositions);

(d) names of items, properties, and relationships (symbolic references which link items to further knowledge representations within the system).

These items can be represented in the form of hierarchical inheritance networks, as will be discussed later. Increasingly specific higher level items are composed from more general lower level items, down to the basic drawing entities used by the drawing machine. These items condition arrangements of parts of drawings which can occur in a current graphical discourse and they are used to interpret drawings.

Logical operations can then be performed on drawings in a similar manner to the way in which they are applied to text expressions. Predicates, as used in predicate logic, can be defined in terms of conditions that need to be satisfied. These conditions can specify the necessary presence of graphic items (in place of words) with specified properties and in specified relationships to other items. Matching drawings to graphical concepts to establish whether conditions are satisfied can be done provided that the production of drawings is conditioned by currently active graphical concepts.

Predicates can refer to abstract definitions of spatial objects, situations, and actions, and they can be incorporated into graphical concepts. When considering an application domain, for example, we can think of *moving* something from 'here' to 'there'. If the thing has a graphical description and if its depiction has a part by which it can be moved, and, similarly, if the origin and destination for the move have graphical descriptions within the same drawing space, then we can expect to describe the move behaviour graphically. The graphical expression will depend on some satisfactory mapping between the *move* predicate as known in the application domain, and a graphic *translate* predicate as known to the drawing machine. For practical applications, many variants of such predicates are likely to be needed, specific to kinds of thing to which they apply, and all necessary mappings have to be specified to a system. In the case of natural language systems, mappings usually are prespecified and they are explicitly linked to entries in a lexicon. In the case of CAD, we want such mappings to be definable by users, by designers.

A WORKED EXAMPLE
So far in this chapter we have explored general concerns which affect the possibility of interplay between formal systems and other human sensitivities. The remainder of this chapter will present a worked example to illustrate a possible direction for further development of CAD systems. The example is based on current work in Edinburgh — it will refer to the MOLE logic modelling system and will outline objectives, strategy, and indicative results.[5] Current work continues to pose fundamental questions which remain poorly understood and which do not have conclusive answers. Thus, this worked example has to be accepted as an illustration of intentions, but these intentions are vital to future practical CAD systems.

Objectives
The initial motivation for MOLE was prompted largely by experience of the Scottish Special Housing Association's housing and site

Designing in Words and Pictures

layout design system, described in Chapter 4. This experience prompted a concern about the prescriptiveness of systems which results in designers becoming dependent on systems which cannot respond to changing demands made on designers. As described in Chapter 3, these demands are made by other people in the world in which designers work.

This concern led to the concept of description systems.[6] The intention for MOLE was that it should support descriptions produced by designers, and that it should represent those descriptions as reflections of designers' knowledge within computers. Descriptions came to be conceived as compositions of expressions, in which parts of expressions are logically related. Relationships refer to interdependencies between expressions, as in, for example, 'this wall' of this house and 'that wall' of some other house. These relationships include: *attachment* of one part to another, so that one thing becomes part of the other; *inheritance* of parts, so that some part of this thing is described by some part of another thing; *instancing*, which allows this thing to change without affecting the other thing; and *copying*, so that the other thing can change without it affecting this thing. Changes to descriptions can then be made by editing expressions. Execution of changes by the system should not be conditioned by any notions of correctness within application domains. A description system has to accept any edits and be responsible only for working out the logical consequences on relationships in any given state of a description.

Such a description system has to meet the following criteria:
(*a*) accept expressions in the form of words and in the form of drawings;
(*b*) support correspondence between these forms of expression;
(*c*) store expressions;
(*d*) permit recall of any current states of stored expressions;
(*e*) give access to system-defined functions for establishing relationships between expressions;
(*f*) accommodate expressions which invoke unforeseen changes to stored expressions.

Apart from (*b*) these criteria might appear unexceptional until we add that the term, expressions, has to be understood in the sense of descriptions as discussed earlier. When we do so we find that we are faced with insuperable problems, and we have to accept certain conditions.

An expression can be made to reflect (but not be the same as) anything a person might have in mind, provided that the form of the expression differentiates between:

those of its parts which have system-defined interpretations, which are intended to activate mechanistic functions within a computer; and

those of its parts which remain interpretable and, therefore, meaningful only to the person and other people.

This distinction has an affinity with the old distinction between programs and data; but here the distinction has to be more severe and remain visible in expressions — otherwise the user would lose control of the prescriptive behaviour of the system.

Any number of expressions can be combined so that any sequential reading of some or all the expressions constitutes a composite description of an object, such as a design object, provided that compositions are defined by:

parts of expression which have system-defined interpretations to relations which point to other parts of expressions, where those relations invoke mechanistic functions within a computer.

Any number of expressions can be combined so that any sequential reading of some or all the expressions describes a task as perceived in a user world, provided that compositions are defined as above, and provided that descriptions of tasks include:

parts of expressions which have system-defined interpretations to replacement operations, where replacement operators invoke mechanistic functions within a computer.

This use of sequences, in the cases of object and task descriptions, is similar to the traditional use of sequential processing; but here sequences are user declared, they can be developed into user-defined hierarchies or networks, and they do not preclude the use of parallel processing. Most importantly, they have to be evident in the composition of expressions — otherwise, once again, the user would lose control of the prescriptive behaviour of the system.

It should be noted that all the conditions outlined above refer to prescriptions which constitute the definition of the computer system. For formal systems, some prescriptiveness is inescapable. The overriding objective for MOLE is to keep its prescriptiveness down to a level that is independent of domain knowledge, and to make all currently active prescriptions visible in any expressions that pass between a person and a computer. By keeping prescriptions visible, the person can see what the computer knows and can, therefore, exercise responsibility in ensuring that expressions correspond with his or her intentions.

This objective carries the implication that a person, as the user of the system, has to know the prescriptiveness of the machine and has to work through its prescriptions to arrive at intended expressions,

Designing in Words and Pictures

such as descriptions of design objects. The person has to know the concepts built into the machine, and these concepts should allow the possibility of unrehearsed expressions.

This objective also carries the implication that the system definition should include limited functionality to avoid obvious or hidden conflict with any user's domain knowledge. In MOLE, system-defined functionality is limited to the logical operations indicated in the conditions outlined above, and to simple arithmetic operations. A person using the system should then be able to build up representations of domain-specific tasks, and the same or like-minded users should be able to call on the system to execute these tasks on different applications.

Machine environment
MOLE is implemented in the Prolog logic programming environment.[7] Prolog is a logic system formed as a computational subset of symbolic logic. It also serves as a general computer programming language, which means that expressions which occur in the language have interpretations in machine operations down to the level of operations on binary bit patterns. Put the other way round, interpretations from machine operations support expressions that are true to the logic system.

The virtue of Prolog is that provided you stay within the bounds of its logic system you do not have to worry too much about how your expressions are mapped to machine operations. Prolog is a rule-based system, built on pattern matching and 'true if' inferencing mechanisms. Its resolution capability employs a depth-first left-to-right search strategy applied to hierarchical tree structures, where these trees are constituted as logically determined arrangements of 'facts' and 'rules'. The internal machine operations do not differ substantially from older systems, such as those that supported the data structuring system outlined in Chapter 4, under system implementation — we are still using the same machines. However, the advantage offered by Prolog (and not only Prolog) is that it gives users declarative rather than procedural access to machine operations.

This virtue of Prolog also has its drawbacks. The separation between computational procedures and the logic system presented to users is not complete. Whether complete separation is possible or even desirable for practical and useful systems remains a moot point. The problem is that declarative access to machine operations can fall foul of the machine's procedural representation of the user's knowledge. Prolog's success in masking procedural states makes it difficult for a user to know what the machine knows. This difficulty occurs, for example, when a user cannot see the current states of search trees invoked by new expressions. This can be a serious difficulty when users have to anticipate the effects of adding further expressions.

Despite this reservation, Prolog does represent an advance in computer technology when compared with previous experience of FORTRAN and Unix/C. It is aimed at easing program development, encouraging increased attention to logic and reducing the effort required to implement programs. This is a worthy bias for any system, given that any practical and useful system will always be subject to demands that it be changed. This general condition applies even to low-level representation systems such as MOLE, and even to Prolog itself.

System strategy
MOLE provides a formalism for expressions which refer to knowledge that might occur in the minds of designers, and the formalism invokes machine functions in computers. The result is logically structured arrangements of words represented in a computer such that the arrangements of words and changes applied to these arrangements reflect what designers know and intend to do with design objects. The words that occur in these arrangements can be viewed as naming the parts of concrete or abstract design objects. Arrangements should be thought of as parts hierarchies and networks forming descriptions, in which parts of descriptions can be related to other parts within the same or across different descriptions, as illustrated in Figure 7.6.

MOLE can be regarded as an extended kind of word-processor which supports logical relationships between words. In general, MOLE expressions include two kinds of notation. Expressions include words that have interpretations only in the minds of users and are known to MOLE only as instances of data types. Expressions also include special characters and these have system-defined interpretations which invoke machine functions, and they are used to set up and change relationships between words. These special characters can be thought of as punctuation, as is used in normal writing (see Chapter 8, Technology of Words).

In addition, MOLE supports drawing as a mode of expression. However, drawings need to be understood as expressions given to MOLE, rather than as part of the definition of MOLE itself. A drawing machine is a user of MOLE, using the symbolic repre-

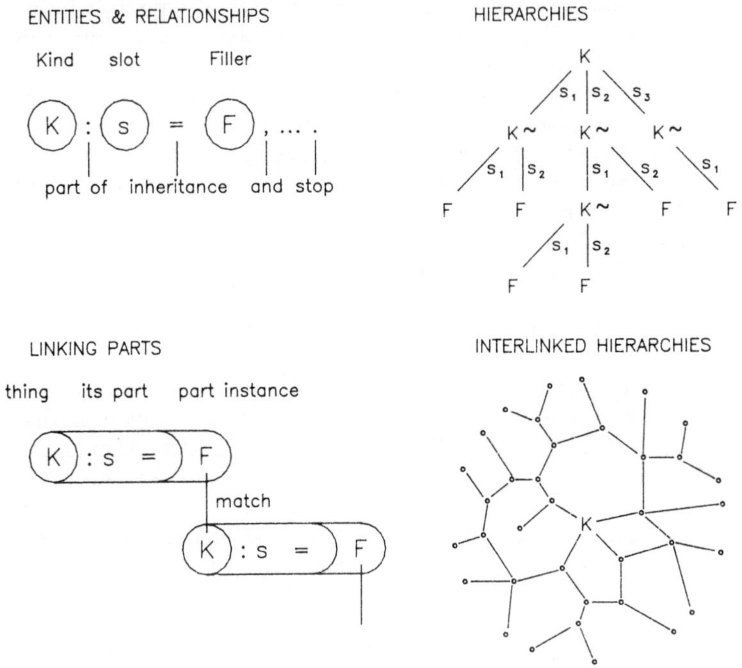

Designing in Words and Pictures

ENTITIES & RELATIONSHIPS

Kind   slot   Filler

$(K) : (S) = (F)$, ...

part of   inheritance   and stop

HIERARCHIES

LINKING PARTS

thing   its part   part instance

$(K) : S = F$

match

$(K) : S = F$

INTERLINKED HIERARCHIES

Figure 7.6 *Parts, hierarchies, and networks.* MOLE's representation scheme employs three main data types expressed as user-declared words: kinds (naming parts), slots (naming parts of parts), and fillers (instructions composed of kinds and slots which identify further parts). Triples of these types are held together by system-defined relationships selected by users (like punctuation). Matching fillers to further kinds results in logical tree structures. Making parts the same as, or new instances of parts in other pre-existing structures results in logical networks across hierarchical structures.

sentation of drawing objects and operations that are held within MOLE. This position will become apparent later.

The MOLE formalism employs kinds, slots, and fillers in a manner resembling frame systems but without any special status for collections of slots as frames.[8] This formalism simply serves to differentiate between words or drawings that name or depict parts of design objects, words that link one part to another, and words that say what other part is being linked. The system sees these distinctions as follows:

A kind (K) is an expression that names or depicts something, any kind of thing a person (or people) might have in mind. A kind also exists as a unique item of data of type K, within a machine. Kinds can be expressed in the form of any user-declared words or pieces of drawing.

A slot (s) is an expression that establishes a 'part of' relationship between one K and another K, so that a sub-K is a 'part of' a super-K or the super-K 'has a part' sub-K. Any slot is known to the machine only as an instance of a 'part of' relationship, which has to be unique in any set of s attached to a super-K.[9] Slots are expressed in the form of any user-declared words or pieces of drawing.

A filler (F) is an expression which denotes a path name which identifies a particular sub-K, within the context of the parts which describe some other known super-K. A path can be any succession of zero or more slots from a super-K down through its sub-Ks. Thus, a path name evaluates to a kind, and a path name which consists of only a kind name evaluates to that kind. A filler can be expressed as any series of kinds and slots linked by special characters that denote system-defined relationships. A filler can also include further data types, such as constants (that are not intended to receive further descriptions) and numbers (that take part in arithmetic functions).

The normal form of a MOLE expression is K:s=F. This says that there exists a kind named K, which 'has a part' (denoted by :) which is named s, which 'is filled by' (=) a path name, F, to some other kind. Any number of slots can be attached to the super-K, and the sub-Ks indicated by their fillers can also be super-Ks with their own slots and fillers. In this way, descriptions can be built up as parts hierarchies, with nodes as fillers referring to Ks and branches as slots. All Ks are potential super-Ks and sub-Ks. A part can be any K plus any further slots and fillers that it might have down a parts hierarchy. This strategy is illustrated in Figure 7.6.

An important feature of MOLE is its inheritance mechanisms which are defined at the level of the representation formalism. These mechanisms allow kinds from some existing context (as parts or sub-Ks of an existing super-K) to be declared as parts of another super-K, without changing the parts or sub-Ks. Thereby a kind can be made to inherit parts in its description from one or more other kinds. Alternatively, inherited parts may be changed in a new context, under a new super-K, so that the changes occur only in the new context. In this case, inherited parts become new instances in the new context, permitting the further parts of inherited parts to be overwritten. Successive inheritance by different super-Ks can occur, and they can include instancing at any step in a succession. Through inheritance, different parts hierarchies can be linked together to form logical networks, with successive parent and child instances of kinds (Figure 7.6). A major effort in developing MOLE has been the definition of inheritance mechanisms which will deal with all possible states of kinds. These mechanisms are invoked by system-defined characters, usually ~ denoting 'an instance', as will be apparent in the example.

MOLE offers a rule-based representation environment. Rules are invoked when the system evaluates a user's surface expressions. The user can also declare rules explicitly within surface expressions, embedded within fillers or as separate expressions, and such rules can refer to parts of many different descriptions. The ability to express rules, and the implications of doing so, are discussed at the end of the following example.

Drawings produced by a drawing machine can be represented symbolically in MOLE. For this purpose, the system has a representation of the general structure of line drawings, as shown in Figure 7.7 and following the general decomposition outlined earlier (Figure 7.4). Construction lines, line segments, chains of connected segments (shapes), and compositions of chains are represented as kinds and are related by slots and fillers. The representation includes transformations which can be applied to any of these parts of drawings, and the effects of transformations are conditioned by current status of angle and length values of parts, in the manner shown in Figure 7.5. Drawing productions automatically get represented as instances of kinds, slots, and fillers which can then also serve as parts of non-graphical descriptions. The resulting interconnection between graphical and textual descriptions of design objects is illustrated in the following example.

Example
With this outline of the representation environment offered by MOLE, plus its representation of drawing objects and operations, we can now illustrate how it is used to make and change descriptions of design objects. The purpose of the example is to convey an impression of what it is like to converse with a computer and to get it to do what you want. Given that computers are machines that are materially dissimilar to people, it should not be expected that talking to computers is easy. In the example, expressions have to identify machine functionalities which are domain independent and

Computer Discipline and Design Practice   198

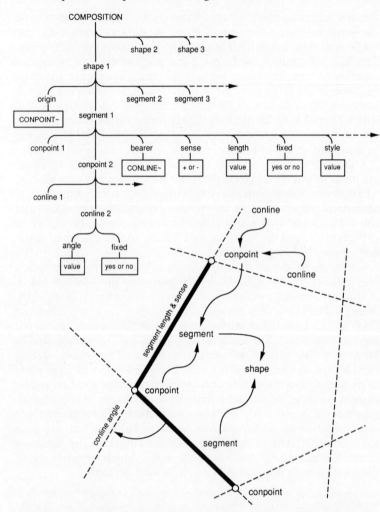

Figure 7.7 *Logical structure of drawings*. The general structure of line drawings is represented in MOLE as a hierarchy of parts of drawings, in terms of kinds, slots, and fillers. This representation sees drawings as composed of construction lines with angle values, construction points at intersections of construction lines, line segments delimited by construction points and with length values, shapes consisting of open or closed chains of segments, and compositions of shapes. This representation is independent of the drawing space used by the drawing machine that uses MOLE (no co-ordinate point data) and it contains enough information for purposes of regenerating drawing productions.

Designing in Words and Pictures   199

invoke those functionalities in the context of other parts of expressions which are domain specific, that relate to a user's own view of the world. Expressions have to cause the machine to behave in a manner that the user will recognise as being appropriate to an application domain.

The following example illustrates a deliberate use of logic to support a particular view of design objects and tasks. It makes no claim to illustrate the best or even a good computer language — it simply illustrates a possible treatment of hierarchical parts descriptions of design objects. It also makes no claim to represent the best or even good design knowledge — it simply illustrates a possible view of an aspect of design.

For purposes of illustration, we can return to the example of walls and wall junctions discussed previously in Chapter 4, and we might want to remove the condition of orthogonality as indicated in Figure 7.8. To keep this illustration brief, we will consider only end-on

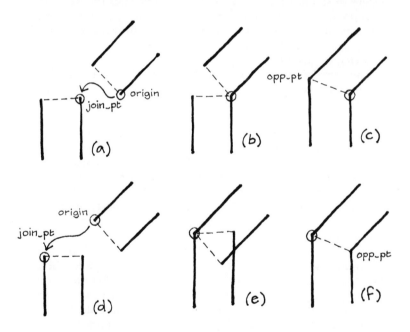

Figure 7.8 *Problem of wall junctions*. Drawings of wall components, initially depicted as stretchable rectangles, need to be modified when walls are brought together to form junctions. Faces and ends have to be stretched and rotated to result in continuity of wall parts for any angle of junction.

junctions and will presume that the system already knows something about walls and drawings of walls. Previously input expressions will have said that drawings of walls start off as variants of rectangles, with long segments that are stretchable and short segments of fixed length. The long segments are referred to by named parts of walls, 'face1' and 'face2', and the short segments are named 'end1' and 'end2'. The system knows that walls have faces and ends and it knows of wall types (external, internal, etc.) which govern the lengths of ends, and wall materials (brick, block, etc.) which can be associated with the drawings of walls. When any two walls are to be joined, the drawing depicting one wall is moved to the drawing depicting another wall, the latter already being fixed within some arrangement of walls. Using a drawing system, one wall is moved by translating a conpoint of its drawing, which becomes that wall's origin. The conpoint of the drawing of the fixed wall, to which the origin of the moving wall is translated, becomes the fixed wall's join point. The two walls then share the same join point which corresponds to a conpoint in the composite drawing of the walls. These steps in forming a junction are illustrated in Figure 7.8(*a*) and (*b*), or (*d*) and (*e*).

The remaining task, which is the subject of this example, is to change the faces and ends of each wall in order to form a continuous junction between the two walls. This goal is indicated in Figure 7.8, in the transition from (*b*) to (*c*), or (*e*) to (*f*), and it is illustrated in more detail in Figure 7.9. The goal is to grab hold of the ends of the wall faces opposite to the join point and move them to the intersection of the underlying construction lines of those faces. In moving these ends of faces, we want to stretch (or contract) the faces until they meet, and the wall ends need to be rotated until they coincide. The intention is to prepare a general description of this task, which can then be executed in any particular context set by any given pair of walls.

The resultant description of the general case of an end-on junction is set out below. It includes its own join point, opposite point, and change parts which refer to other parts that will be found in any given pair of joined walls. The change parts include move parts which effect changes to the relevant wall faces and ends. The composite expression employs MOLE's conventional notation, with kinds in upper case characters, and slots in lower case characters — the interpretations for the parts of the expression will be explained in the subsequent paragraphs.

Designing in Words and Pictures

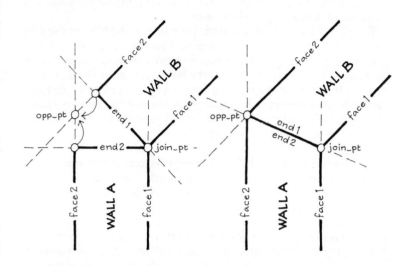

Figure 7.9 *Joining walls*. A general description of end-on wall junction, as represented in MOLE, is depicted (construction lines are shown dotted and segments solid). The logical description has to identify those parts of the adjoining walls which are parts of the junction, to arrive at the junction's join point and opposite point. An intersection point in each of the wall drawings has to be translated to the junction's opposite point, dragging face and end segments along with it.

    JUNC_END:
      [join_pt = fix_wall:join_pt,
      opp_pt:
        [conline1 = fix_wall:face*|{:conpt* = join_pt}:bearer,
        conline2 = join_wall:face*|{:conpt* = join_pt}:bearer],
      change:
        [fix_end = fix_wall:end*{:conpt* = join_pt},
        join_end = join_wall:end*{:conpt* = join_pt},
        move1 = fix_end:conpt*|{ = join_pt}:(opp_pt:* = *),
        move2 = join_end:conpt*|{ = join_pt}:(opp_pt:* = *)]].

JUNC_END is a kind name declared by the user, for this kind of junction. The common-sense translation of each part of this composite expression is as follows.
    [join_pt = fix_wall:join_pt,
First, the parts needed to form the continuous junction are identified. The first part is the junction's *join point*, which is (=) the part that fills the join point of some fixed wall. From the previously

given information, the system knows that this join point of a wall is a conpoint of a drawing of some wall.

The part of the expression to the right of = is the filler of the *join point* slot of the junction description. This filler is expressed as a path name consisting of a succession of slots. This path starts from the nearest kind which possesses the path's left-most slot as one of its slots. In the present case, that kind is not as yet identifiable from within this description — it simply refers to some fixed wall part that will set the context for some instance of junction (as will become apparent at the end of this example).

opp_pt:
 [conline1 = fix_wall:face*|{:conpt* = join_pt}:bearer,

The next part of the junction is its *opposite point*. This part has its own further parts. One of these further parts is the *opposite point's conline1*, which is the part that fills the bearer (a conline underlying a segment) of some face of a fixed wall. Again from previous information, the system knows that this conline is a part of a drawing segment that describes a face of some wall. Limitations on which face this might be are identified by the conditions expressed between curly brackets. These conditions state that the face is any face (*) that does *not* include among its parts (|{:) any conpoint part (of the corresponding drawing segment) that *is* the join point part of the fixed wall. This exclusion condition, in the present case, is effective because any wall has previously been defined as having only two faces.

The opposite point's *conline2* is described in similar fashion, and the joining wall is as yet unidentified (like the fixed wall, above).

change:
 [fix_end = fix_wall:end*{:conpt* = join_pt}],

Here the parts of the junction which need to be changed to form the continuous junction are identified. The change parts include a *fixed end* part, which is some end of a fixed wall, being a segment in a wall drawing. The conditions that identify the required wall end state that it is any end of a fixed wall (a drawing segment) that *does* include among its parts ({:) any conpoint (a drawing conpoint) that *is* the fixed wall's join point.

The change part's *joining end* part is described in similar fashion.

move1 = fix_end:conpt*|{ = join_pt}:(opp_pt:* = *),

Finally, the intersections of the wall ends and faces which are to be moved to the junction's opposite point are identified and moved. The change parts include a *move1* part, which is any conpoint of the fixed end (a drawing segment) that is not the junction's join point (a drawing conpoint). This *move1* part is then described further. Its parts (its conlines) are now made to include all the parts of the junction's opposite point. What is happening here is that the description (the slots and fillers) of one part of the junction somewhere within the composite expression, gets added to (and overwrites corresponding slots in) the description of another part of the junction. This replacement operation is invoked by the round brackets at the end of the filler to the junction's *move1* part. The effect, in this case, is that the conlines which describe the fixed end's conpoint are replaced by the conlines of the opposite point. Thus, this conpoint is translated to the junction's opposite point. Anything else described by the same conpoint, being all parts (super-Ks) described by that conpoint (the sub-K), will also change. Thus, the move results in changes to the drawing segments that describe a face and end of a fixed wall, as shown in Figure 7.9.

The change part's *move2* has a similar effect in changing the drawing segments which describe a face and end of a joining wall.

Note that system-defined change or replacement operators are used only in the last two steps in this description of an end-on junction, after all its parts that will be affected by the change have been related to existing parts of the adjoining walls. Remember, also, that only the special characters have system-defined interpretations, calling on system functionalities. All the words (or word abbreviations) are known to the system only as kinds (in this case, only the initial name) and slots. The fillers to the slots indicate further kinds. By evaluating fillers, the system should exhibit behaviour as though it knows about wall junctions, when it evaluates this description.

So far this general description of end-on junction cannot be evaluated. Before this can be done, the description has to be set in a context of two adjoining walls. This is necessary to identify the parts of those walls referred to by parts of the junction: the parts of the fixed wall and joining wall. An instance of junction with the adjoining walls can be expressed as follows.

JOIN_END_AB:
  [fix_wall = WALL_A,
  join_wall = WALL_B,
  join = ˜JUNC_END].

Now we have two particular walls, A and B, and their continuous junction is formed by adding this junction's own instance (˜) of the general description of end-on junction. The system knows that the junction inherits its own instance of the general description.

JOIN_END_AB:
[fix_wall = WALL_A,
join_wall = WALL_B,
join:
  [join_pt = fix_wall:join_pt,
  opp_pt:
    [conline1 = fix_wall:face*|{:conpt* = join_pt}:bearer,
    conline1 = join_wall:face*|{:conpt* = join_pt}:bearer],
  change:
    [fix_end = fix_wall:end*{:conpt* = join_pt},
    join_end = join_wall:end*{:conpt* = join_pt},
    move1 = fix_end:conpt*|{ − join_pt}:(opp_pt:* = *),
    move2 = join_end:conpt*|{ = join_pt}:(opp_pt:* = *)]]].

The system is able to execute this instance of junction by evaluating the fillers of junction parts to parts that will be found in the further descriptions of walls A and B. Evaluations result in filler expressions being replaced by the particular kinds which they identify, which means that the description gets changed. However, it is only this instance which gets changed, without affecting the general description of end-on junction. Furthermore, since evaluated kinds include parts of drawings, execution of this instance of junction results in changes to parts as seen by a drawing machine. Thus, the newly formed junction represented within MOLE can also be displayed in the drawing machine's depiction of the

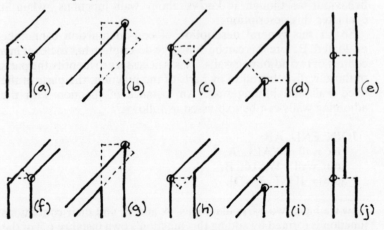

Figure 7.10 *Junction configurations*. The general description of end-on junction can serve any configuration of adjoining walls, rotated to any angle and covering walls of unequal thickness, except for (j).

junction. Used in this way, this general description of end-on junctions will serve all instances of such junctions for walls at any angles and of different thicknesses, as shown in Figure 7.10.

Rules
It can be noted here that the whole of the composite expression describing the general case of end-on junctions can be regarded as a 'true if' inference rule: 'we have an end-on junction (initial kind name) if all the conditions (the parts of the expression after the kind name) are satisfied'. Alternatively, the expression can also be read as an 'if true then' production rule: 'the rule named end-on junction says — if all the conditions (up to the last two steps) are satisfied then execute the action (the last two steps)'.

These are plausible and logically equivalent readings from outside the system, even if there are no obvious and explicit rules within the composite expression. Consider the visible conditional expressions within the example which are used to restrict solution spaces and which occur in fillers expressed as path names. These conditional expressions subsume a 'true if' rule in the following form: $s\{...\}$, which says that s is true (for the path denoted by the filler) if the conditions between curly brackets can be satisfied. Conditions can refer to anything within some context, including any part or parts that may be found outside the current description ($\{:s_1 = s_2\}$), and they can specify a part that has to be matched by the part that fills the slot at the head of the rule ($s_1\{=s_2\}$). In anticipation of possible failure, an or (;) can be added to the rule to indicate an alternative path from the rule ($s_1\{...\};s_2$). Thus, the system employs rules when evaluating expressions and the user can also include rules explicitly within expressions.

As with any expressions that are composed of, or imply, rules any MOLE expression incurs a risk of failure. A rule will fail if the conditions it specifies cannot be found from within the composite expression in which the rule occurs. Responsibility for avoiding failure has to rest with the person who uses the system, when formulating expressions to the system. This is an awesome responsibility when an expression invokes many different rules. This responsibility can be reduced only by prepackaging bodies of information, by someone else, so that those bounded and securely wrapped packages can be safely called by a user. Such a strategy would raise, once again, the principle issue of prescriptiveness of systems.

Conclusions

The example shows a user-declared description of a task, which remains available for reuse in different applications. The task is executed by the system reading and evaluating the parts of the description sequentially.[10] Evaluation involves finding the kinds indicated by the fillers and their embedded conditions plus replacement operations (expressed between curly and round brackets). Evaluations result in logically structured relationships between kinds within and across different descriptions which correspond to logical hierarchies and networks (Figure 7.6).[11] The structures remain visible in expressions by means of the slot names that denote which part of one description forms which part of another description, and the system keeps account of inheritance, instancing, and copying of parts in descriptions. By reading and evaluating sequences of parts in a composite expression, the system is able to show the logical consequence of what the user has said.

The system is able to perform logical evaluations of descriptions, despite having no interpretations for any of the words used in a description. The system recognises only that words are kinds or slots, and keeps track of different instances of kinds and slots. All words have meaningful interpretations only in the minds of people who use the system.

The system's functionality is invoked only by system-defined characters (the punctuations and brackets) which effect relationships between words. A user has to know and work through the prescriptions invoked by these characters in order to control the operations of the machine. By executing these operations, the system behaves as though it shares the user's understanding of what it is required to do.

Thus, we have a system for making and changing descriptions. The system knows nothing about what is being described, yet it can be made to behave in a manner that appears meaningful to a user. Only the user's knowledge is being represented by descriptions and the user can cause the system to reflect changes to that knowledge in the changes it executes on descriptions.

The example also demonstrates a correspondence between symbolic expressions of knowledge about walls, in the form of words, and drawn expressions of that same knowledge. For this purpose, drawings are surface expressions produced by a drawing machine that employs the MOLE representation environment. The junction description refers to parts of drawings, such as faces and ends (line segments), intersection points, and construction lines, but the user does not need to know particular instances of these parts as known by the drawing machine. In the example, the user needs to know only that a drawing of each of any adjoining walls includes two faces and two ends, and either a join point or an origin. When a replacement operation is applied to a conpoint of a drawing, its effect is conditioned by the status of length and angle values of the adjoining line segments — as part of the specification of the point translation (not shown in the example) invoked by this operation.

This correspondence between words and drawings rests on a common underlying representation scheme for both forms of expression. The importance of this correspondence is that it allows access to parts of drawings and supports changes to those parts by means of words used to describe the objects that the drawings depict. The user does not need to know and work with a separate representational structure for drawings, detached from a representation of depicted things. Instead, the user can cause changes to any parts of drawings by changing the words that describe depicted things. Here we have the promise of a new integration between thinking about and drawing design objects.

Returning to the three questions posed at the start of this chapter, we can now arrive at tentative answers. First, a computer's representation of domain knowledge can be made to evolve with designers' changing practices, provided that designers can see and control the effects of low-level prescriptions that define the system. Secondly, given this degree of control by designers, computers do not need to know their users (beyond recognising their identity), nor can they do so. Lastly, it does seem possible to conceive systems that can support the link between drawing and thinking about design objects. However, the worked example may well prompt the conclusion that using computers in this way, by having to exercise control of low-level operations of a machine, is not easy. We then need to recognise that this condition is unavoidable in any use of computers, and it has to be faced by users or by other people who prepare computer applications on their behalf. This condition can be masked from users only by resorting to higher level prescriptions which will intrude upon people's domain knowledge.

We need to discover more about the possibility of linking symbolic and drawn expressions to represent people's knowledge about design objects, to reflect changes to that knowledge, and to reflect differences between separate knowledge of individual designers. Potentially, the system strategy that has been illustrated might overcome the limitations presented by previous experience of

CAD (Chapters 4, 5, and 6). However, this strategy implies some future acceptance of precise logical constraints on expressions, which implies an acceptance of computer literacy on a par with our more familiar word literacy. These implications will be considered further in the next chapter.

SUMMARY

This chapter has set out a strategy for CAD, illustrated by the example of the MOLE logic modelling system. The ambition is to devise computer systems which can support any expressions from designers about design objects. They have to support unrehearsed expressions in the form of words and drawings which describe design objects. We want description systems.

A description system is one which supports the production and modification of expressions which describe design objects without any prescriptive functionality for determining the correctness of descriptions with respect to anything outside the system. The system is responsible only for maintaining logical relationships between parts of expressions and, when changes are declared by a user, the system's behaviour in re-establishing relationships can look like it is executing user-specified tasks. The system does not need to know what expressions mean, in the sense that users know their expressions. The purpose of descriptions is to serve as expressions passing between people.

In MOLE, this strategy accommodates expressions which consist of words declared by users, which have interpretations only in the minds of users and which are known to the system only as differentiated instances of words. Expressions are articulated by using special characters (like punctuations) which invoke machine functions to establish logical relationships represented in hierarchical inheritance networks. These functions include changing or replacing words under user-declared conditions. The same or like-minded users can then get the system to recall past expressions and employ them in further descriptions, to build up representations of their own design knowledge.

Drawing, as a mode of expressions used by designers, presents a spatial modelling environment that can be represented in a logic modelling system. A system which is used to construct and manipulate drawings becomes a user of the logic modelling system. Parts of drawings, relationships between these parts, and transformations applied to parts result in corresponding logical representations in the same form as that which results from text expressions. Interrelated graphical and textual expressions can then be used to

Designing in Words and Pictures 209

describe design objects and tasks.

A general structure of drawings has been outlines, defined in terms of lines, intersections, and kinds of attachment. This structure can be used to describe spatial properties and relations employed in people's graphical concepts. Thus, drawings can be used to depict situations such as things being part of, next to, and near other things, and depict things being moved along or to other things. In CAD, we then have to face a further general condition. When a designer uses a computer system, the designer's purpose is not to gain access to a given model of an application domain. The primary task of designers is to design models that cannot otherwise be deduced from pre-existing models. This carries the implication that designers have to employ domain context free concepts for purposes of describing context sensitive concepts. Here we have a major and speculative ambition: a description system must allow descriptions of new concepts.

Notes to Chapter 7
1. Following on the discussion in Chapter 2 — if we were to think of expressions being meaningful to computers, then, given the different kind of thing that computers are, we could not share their sense of meaning.
2. A generalised logical representation of drawing structure and transformations has been proposed by Szalapaj (1988; *et al.*, 1985).
3. GKS (Krishnamurti and Sykes, 1986) is an example of a general purpose graphical production machine, but its definition has grown from a focus on outmoded computerised production techniques (fast picture plotting on old DVST displays), with little regard to generalised techniques for interpretation aimed at user modules.
4. Work on such links, in the context of formal treatments of graphics and natural language, is described by Pineda *et al.*, (1988).
5. The worked example refers to the MOLE logic modelling system and associated logical representation of drawings (Bijl and Szalapaj, 1984; Krishnamurti, 1986; Tweed and Bijl, 1988), developed at EdCAAD, University of Edinburgh, and supported by the UK Science and Engineering Research Council.
6. EdCAAD's early definition of description systems is presented in its report on Integrated CAAD Systems (Bijl *et al.*, 1979).
7. For a full description of Prolog see Clocksin and Mellish (1981), and on logic for problem solving see Kowalski (1979).
8. Frame systems are developed from the work of Minsky (1975).
9. The use of slots in MOLE has some affinity with semantic networks (Woods, 1975), but MOLE does not attach any significance to names of arcs.
10. In the worked example, the condition that a reading must be *sequential*

is not necessary since, once all the facts have been expressed to MOLE, the system will be able to evaluate any part (any K:s = F) in any order by following the references through slots to any other parts. It would be possible, for example, to ask the system to evaluate:

JUNC_END:move1 = fix_end:conpt*|{ = join_pt}:(opp_pt:* = *)

without having to restate anything about other parts. This means that the system does not need to store information about the ordering of parts. However, this remains a speculative claim that has not been tested for more general cases.

11. Hierarchical inheritance networks in MOLE emanate from users, whether they be humans or other systems like a drawing machine. Thus, in principle and from MOLE's point of view, the kinds it represents are distinguishable as parts of particular users. Kind names rarely need to be included in expressions from users and, in the worked example, the description of end-on junction could have commenced with a slot name:

junc_end:
 [join_end = . . .].

The system would know that this is a junction of John, if it is John's description, and this description refers to parts of descriptions from the drawing machine. This absence of kind names eases dialogue between a user and the system — the user does not need to know the kind instance names known to the system. A dialogue can be conducted without the things being talked about, the kinds, being present in the dialogue — a condition echoed in normal human dialogue.

CHAPTER EIGHT

OUR FUTURE

*Now come all of you workers
who work night and day,
by hand and by brain
to earn your pay,
who for centuries long past
for no more than your bread,
who fought for your country
and counted your dead.
In factories and mills
in the shipyards and mines,
you've often been told lads
keep up with the times,
for your skills are not needed
they've streamlined the job,
with stopwatch and slide rule
your pride they have robbed.
But when the sky darkens
and the prospect is war,
who's given the gun
and then pushed to the fore,
and expected to die
for the land of his birth,
when he's never owned
one handful of earth.
He's the first one to starve
he's the first one to die,
he's the first one in line for
that pie in the sky,
and he's always the last
when the cream is shared out,
for the worker is working
when the fat cart's about.*

> *It's all of these things*
> *that the worker has done,*
> *from tilling the fields*
> *to carrying a gun,*
> *yoked to the plough*
> *since time first began,*
> *and always he's expected*
> *to carry the can.*
>
> *Workers Song*
> *Ed Pickford*

Here we have a cry of anguish from those who see themselves as disadvantaged. How far does that group extend, can it include designers, and who are the oppressors? Is it possible that we are our own oppressors? Can we reshape our world for a better future?

The past seven chapters have explored ideas about knowledge and techniques by which we seek to externalise human knowledge, the assumptions underlying computer technology, and an understanding of design. To draw these discussions to a close, I will now make some general observations about our acceptance of technology, illustrated by the example of our use of writing. I choose this example because it is so widely accepted as obviously beneficial, but I suggest that it has also led to some unattractive consequences. I do so not to decry word literacy, but to indicate the kind of consequences that might come from widespread acceptance of computer literacy. Awareness of these possibilities can motivate advances in computer technology that are compatible with all people's interests. I will touch on broad considerations extending into ordinary activities of people, politics, and survival. I will be speculative in revealing my personal opinions and beliefs, unbounded by considerations of formal correctness, speaking as a person to people.

CHOICES

Central to all the previous discussions is the distinction between shared overt expressions of knowledge and individual, intuitive knowledge. It is not possible to give a precise overt explanation of this distinction; the definition of intuition precludes us from doing so. This distinction becomes important when it is linked to a further notion: human knowledge is evolutionary. Again, we do not have, and do not need, a formally correct explanation of evolution. It is enough to observe that states of knowledge change over time, within persons and collectively among large numbers of people, and that changes are not additive and do not form a linear progression to some perceived ultimate goal. By labelling what we observe as evolution, we imply a link between the ability to change and our interest in human survival. This reference to survival is deliberate. It then appears that intuition provides a necessary contribution to evolution of knowledge. People have the ability to employ their own intuitions and this is necessary to our survival as a species and as subgroups of the species — as artists, craftsmen, and scientists, and as businessmen, administrators, and politicians.

What, then, are the implications of new information technology? We can regard all instances of technology as encapsulating certain states of knowledge. Somehow, we think we know what we want to do, we get to know how to do it, and then we know how to construct devices that will do it. All this has to happen in an overt world, prompted by intuition. When we change our mind about what we want to do, or we discover that a technology does more than we intended, we then have to unravel and recover our knowledge from the technology, and try again. Looking back on our past, this unravelling is not something we are good at doing, but we may have to learn.

There are two important considerations that bear on this view of technology. First, there is a separation between the knowledge of people who initiate and develop a new technology, and the knowledge of other people who are invited to use the results. There can be, and often is, a mismatch across this divide between these separate states of knowledge. One reason for this mismatch is that the initiators tend to work in accordance with overtly defined worlds, whereas users working within loosely defined worlds tend to rely more on their intuitions. Thus, we get the familiar distinction between the orderly (and therefore respectable) worlds of researchers and the messy worlds of applications, as well as the counter accusation that the abstract worlds of researchers are unreal when compared with the reality of practical worlds.

Secondly, the orderliness of overtly defined worlds offers the promise of predictability, inviting an alliance with commercial and political interests. These alliances are interested in influencing and exercising power over people. If this association appears far fetched, consider the recent transition in the field of AI from poorly funded exploratory research to highly funded commitments to produce useful results, linked to industry and backed by political ambitions targeted at the international world. Such expressions of faith and subsequent major investments in technology have the effect of increasing the separation between kinds of knowledge in different people. Commitment to make a technology work and suppression of doubts about whether it can work or whether we want it to work mean that the knowledge encapsulated in the technology will become increasingly estranged from the knowledge of other people. All people then have to live in the world conditioned by that technology.

These observations point to a worrying prospect. We can foresee a world that will be divided between those people who have knowledge and control of technology, and thereby have power over other people, and those many other people who do not possess that knowledge and, therefore, have little power to influence the circumstances in which they live. This suggests a new kind of class division which might take the form of the Third World phenomenon translated into the developed world, producing a class of powerless and acquiescent people within developed societies. Again, if this prospect appears far fetched, consider our experience of nuclear technology, covering energy and weaponry, and the division it has produced between people who are equipped to know and those many other people who are deemed unfit to know, locally, nationally, and internationally. In a less dramatic but more all-pervading manner, we are experiencing the same pattern with respect to techniques in general — we are expected to value techniques as things that work without regard to what they represent and imply for people. The divisiveness this induces may already be with us; it was identified by sociologists of the 1960s as 'alienation'.

The problem is not one of improving the lot of certain groups or classes of people relevant to other people. It is potentially much more serious. The problem refers to our intellectual and political mechanisms for change, affecting our ability to let knowledge evolve. These mechanisms can inhibit the self-correcting behaviour essential to survival. Here I mean survival of all life and I am referring not only to the possibility of a nuclear holocaust. In quiet ways, progressive disruption of the earth's ecology by human industry could do the job. It is to be hoped that this is an over-dramatisation. But, and here is what I believe to be the crucial point, to ensure that this will prove to be an over-dramatisation we have to place faith in people. All people, individually, need to retain the ability to contribute to our collective survival. We must, therefore, ensure that our machinery for externalising knowledge will admit intuitive, holistic contributions. The issue presented by technology is not: can we make it work? Instead, the question should be: can we conceive of a technology that does not automatically increase the ascendancy of overt knowledge, that can make room for intuition?

Two faces of technology

Of course, the benefits of a technology are persuasive when it enables us to do more things more quickly and with less effort, if we want those things. Mechanical devices provide good examples. In agriculture, development of the tractor enabled fewer people to turn more land, and more so as tractors got bigger. It was not intended that tractors should also chase all kinds of people off the land, nor that land should be transformed into ever larger unbroken parcels, but that has been an effect of technology. Such effects, multiplied by different technological advances, have conditioned our concept of efficiency. We have come to believe that it is always better that fewer and fewer people should be able to do more and more. Unfortunately, the people who are no longer needed do not disappear, upsetting the economics of efficiency. It is possible to conceive an opposite notion of efficiency: more and more people should gain a livelihood from a given activity. Could such an ideology be compatible with technological progress? In the example of agriculture, the measure of efficiency would then be the largeness of the population earning a livelihood from working the land, without loss to client markets. Our technological achievements now inhibit our ability to think in this way.

We need to consider similar effects of new technologies aimed at extending intellectual abilities. To do so, we have to look anew at how we employ these abilities in established fields before computers are introduced. Thus, in Chapter 2, we considered the thinking ability of people and the role of language. Later chapters then applied this understanding to design, to the role of theory in design, and to examples of practical design applications. How do we intend computers to affect design, the role of designers and the expectations that other people have of design products? Are we seeking to apply some notion of efficiency that expects fewer designers to do more design, or are we striving for better designs? Can we make the use of computers compatible with designers' intuitions and innovation?

The view I have advocated is that advances in computer technology must preserve people's role in design as we know it. To do anything else, to change or replace design simply in order to exploit current technology, would be an arrogant negation of human evolution. The criterion for new advances must be that computers can reflect any unrehearsed thoughts in the minds of designers, in response to what designers tell them, and that computers must enable designers freely to access and manipulate stored expressions of their design knowledge. We can find theoretically sound arguments which tell us that it is impossible to meet this criterion. Any overt and formal representation can be only a filtered representation of something in the mind of a person. None the less, this criterion is valid in motivating those advances that will be most acceptable to designers. We want to aim for a computer technology that is no more restrictive than the familiar representation environments we employ when writing and drawing.

Motivation for designers
As computer technology becomes an increasingly established means for externalising and processing people's thoughts, alongside and perhaps in place of words printed on paper, so clients and producers will 'speak' to designers through their computers. For the role of design to survive in such a world, designers will have to become computer literate, to articulate overt expressions of their thought. The option of withdrawing from technology, away from computers and into their own minds, will not be tenable for people who have to make their responses evident to other people. To preserve design as a valued human activity, designers need to become familiar with, and take part in, the development of this technology. They need to develop their own deep awareness of values implicit in the technology and should demand advances that will be responsive to people's design practices.

Thus, the motivation for designers to use computers is not the benefits they may obtain from single instances of computer applications. Anticipation of such benefits may influence immediate decisions of particular designers, but motivation survives even when particular anticipated benefits are not realised. The broad and persistent motivation is our general acceptance that computer technology now forms a natural part of human development. Computer technology is now so firmly established that people, in general, are coming to regard this technology as good and necessary, in the manner that we view word literacy unquestioningly as being good.

TECHNOLOGY OF WORDS
To gain some insight into the possible future effects of computer

literacy, we can re-examine our familiar word literacy and the effects of our technology for representing words on paper. The following paragraphs build on the discussion of language in Chapter 2, and they will focus on the production of expressions in the form of written words. This familiar form of expression has a structure; we have to work within that structure to produce expressions, and the structure influences what these expressions can represent. Our predominant use of this form of expression then conditions what we can think and, of course, it supports our ability to share these thoughts among people. In developing this view of word literacy, my purpose is to show how a familiar and accepted technology presents the same issues as are now presented anew by computers. I will suggest how our technology of words has had broad ranging effects on the world of people, as an indication of the kind of effects we might expect from computer literacy. Finally, this discussion will illustrate the general point that technology has to be accepted as part of human existence, but that technology must be made to be responsive to changing demands of people. This last point is particularly problematical in the cases of both word and computer literacy, since each can condition people's ability to express new demands.

To think of word literacy as a kind of technology might appear strange. Such a response serves as a measure of our acceptance of word literacy, our taking it for granted. It is worth looking at literacy precisely because it is generally taken for granted and is considered to be unquestionably good.

Structure of written objects
Writing can be considered as a form of expression that is evident as marks on paper. These marks are conditioned by rules for their arrangement, including the formation of characters and relationships between characters. These rules have to be satisfied, irrespective of what particular arrangements of marks are intended to express. We then have techniques for producing arrangements of marks to form written words and word strings. These techniques ensure that the rules governing arrangements are satisfied, and when they are exercised by people we get a writing system.

Written expressions consist of objects and relations that are detached from people, and they include parts of expressions unique to writing. They include parts that do not appear in translations to corresponding expressions in other forms of realisation, in media such as speech. Yet these parts of writing are important to such translations and to further interpretations into abstractions that people might have in mind.

Our Future

The structure of written expressions consists of characters as a type of object, and instances (the members of an alphabet) which are differentiated by arrangements of lines, graphically. Characters are linked by a single uni-directional 'next to' relationship which governs aggregation of instances to represent words. Aggregations are delimited and related to each other by blank spaces which denote a uni-directional 'separate from' relationship for words. Special symbols are introduced which have meanings like *and* and *stop*, denoted by ',' and '.'. These symbols serve as punctuation for purposes of forming compositions of character aggregations or words. Writing systems can then employ further relations, such as 'similarity', to match similar instances of partial aggregations or compositions. This general structure for written expressions is shown in Figure 8.1.

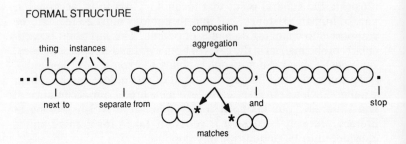

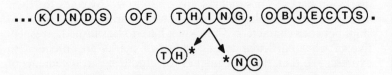

Figure 8.1 *Structure of written expressions*. The structure consists of characters and instances of characters held together by a 'next to' relationship. Aggregations, which we recognise as words, are delimited by a 'separate from' relationship. Punctuations identify further relationships between parts of compositions. Words that are realised in writing then receive interpretations in human language.

This structure has to be known and employed by people who express themselves in writing and who read such expressions. They have to know how to employ relationships between parts of writing

Computer Discipline and Design Practice 219

in order to represent whatever it is that they might be expressing, or to interpret written expressions. This ability has to be learned (typically by each person, during school years); it applies uniquely to writing as a form of expression, and it is an essential pre-requisite to word literacy.

A virtue of a writing system is that it rests on a formal structure which consists of few entities and relations (at least for broad groups of verbal languages). We have learned to use this structure to construct representations of a vast range of human knowledge, from scientific methodologies and problem solving techniques to the rich ambiguities and contradictions of evocative poetry. From earlier discussions, in Chapters 2 and 7, we should recall that in all these cases representations are partial. Thus, we have to regard all expressions as being, to some degree, evocative. Literacy requires the ability of people to construct written expressions that evoke responses within other people.

We can now make two initial observations on our use of literacy. First, people commonly behave as though written expressions can stand for human knowledge, as though expressions in this form are knowledge. People do so despite the inability of any writing to know what it might be representing, or whether it constitutes a 'correct' representation. We have the familiar phenomenon of people calling on the authority of things expressed in writing 'in black and white'. Secondly, the structure which governs the production of written expressions is dominated by uni-directionality, to an extent that is more exclusive than the uni-directional time sequence in spoken language. One thing has to follow on another (including bracketed expressions). All interpretations have to be reconciled with this unbroken linearity. If we attach significance to any interruption to uni-directionality in writing, we effectively move from writing to graphics. Examples of such interrupted writing are found in commercial advertising and poetry.

This belief that written expressions can be knowledge, coupled with the linearity of the form of these expressions, can be expected to have an effect on knowledge within and behaviour of literate people. In the next paragraphs I will suggest some possible connections between literacy and people's knowledge.

Consequences of writing

For written words to serve as the primary reference for any growing body of shared knowledge, we have to accept that written words can convey human thought without the authors of the words being present. We have to accept the separation between intuitive understanding within persons and overt expressions of knowledge on paper, and we have to believe that this separation does not matter. This acceptance has fostered a general notion that all arguments and the correctness of conclusions can exist objectively, detached from persons. Objectivity here needs to be understood as self-consistent linear sequences of expressions rather than the metaphysical question of objective existence. Detachment refers to the evident separation of written expressions from people's intuitions rather than the involved position presented by spoken language. In a literate society it has become imperative that arguments presented on paper, that are intended to survive when passed between many people, must be objective. In this sense, objectivity of knowledge can be viewed as a consequence of literacy.

The thinking processes that people employ have to conform to the criteria for objectivity if their efforts are to be verified by reference to any overt body of knowledge. The principal criterion is that new contributions to knowledge must be consistent with written records of experience of all people, or a sufficient number of people whose experience is acknowledged by other people. The point here is that records, in written form, are inclined to be viewed as imposing their own authority independently of subsequent human experience. Individual thought which deviates from established overt knowledge has come to be the modern equivalent of heresy.

As knowledge represented in writing increases, it becomes increasingly impractical for all people to participate in establishing the validity of every new contribution. We get the familiar proliferation of specialised intellectual disciplines and professional practices. Even in literate societies, specialisation has the consequence that whole fields of knowledge become remote from the majority of people. Even literate people must resort to faith in the specialised abilities of other people. We are not so far removed from the kind of specialisations in the past which fostered belief in magic, but now literacy supports larger numbers of 'magicians'.

Reliance on objective validation is less compatible with those intellectual endeavours which seek to establish links across different fields of knowledge, in response to a broad and involved view of human concerns. Here we refer to knowledge which has to be visible among all people, and we get the justifiable inconsistency of politicians. Knowledge based on objective facts needs to be tempered by wisdom which accepts the moderating influence of people's intuitions.

It might be argued that written words, and the system of writing which we recognise as literacy, is not unique in producing the effects I have outlined. Any externalised recording medium which serves as memory might produce similar effects. What differentiates literacy is the persistence of written words, detached from people, and the sheer quantity of words that exists as records of human knowledge. It can then be said that this observation points to the success of words. However, when people come to regard the records as knowledge, and when people believe that new experience has to be verified by reference to past records, then the records appear to acquire authority over the actions of people, without the records being answerable to people. We are then in danger of acquiescing to a kind of intellectual inertia.

Social structure and authority
We can explore the effects of literacy a bit further. Words appear on paper as inanimate objects, and paper as the medium for holding words is also inanimate. The process of determining which words should appear and the process of associating meanings with compositions of words, of writing and reading, requires human intervention. The representation environment realised in the form of written expressions, embracing the general structure of such expressions and the meanings linked to words, all has to be learned and exercised by people. Thus, we can say that words do not work on paper, they work in the minds of people.

This dependence of written words on people makes the processing of information held in the form of writing a highly labour-intensive occupation. People have to process words, interpret them, and rewrite them for other people, repeatedly for successive other people. If we couple this observation with the so-called information explosion, itself a symptom of many people's attempts to protect themselves by involving ever more people in recording decisions, we can explain the current predominance of bureaucratic information-processing 'machines'. These machines consist of people. When the amount of information becomes very large, touching on different interests of very many people, then the processing of information requires lots of people over long periods of time. Consider the example of building regulations, where the time between a new need being recognised and the corresponding change to a regulation being in force used to be as long as ten years. When information grows to this scale, then it is only very large centralised bureaucracies that can afford the armies of people required to process words on paper.

If we then link literacy with notions of overt knowledge and objective facts, we can see how centralised access to information leads to authority being vested in centralised bureaucracies. Thus, we have experienced the growth of central government and commercial multi-national conglomerates. Whatever other interests also motivated these developments, they could not have happened if we were not living in a word literate society. Once started, these developments appear to be self-perpetuating. They gain a momentum that resists any attempt to reverse the trend towards bigger and more centralised organisations. An all-pervasive inertia sets in, and we find ourselves in danger of losing the spontaneous self-correcting behaviour traditionally associated with a democratic society.

Responsibility
Let us not brush aside these observations as being fanciful. Effects are experienced by designers, through official design guides, regulations, and control of funding directed at selected classes of design objects. In the case of architects, effects are exacerbated by the trend towards litigation (itself an expression of belief in correct answers, rather than in individual and trained experience) and the consequent need to conduct design practices within legally safe bounds. In the face of these developments, we may well worry about a designer's ability to respond to client needs, to be responsible to other people, and to give physical expression to human aspirations in a changing world. We should worry about the diminishing solution space for design objects and the consequent devaluation of design in society.

Word literacy has been described here as a kind of technology in order to illustrate the kind of consequences that any technology can present to people. Even in the example of literacy not all consequences are desirable. Once a technology is devised and mastered by people, it then becomes difficult to control undesirable consequences. Typically, problems presented by a technology are not amenable to technical solutions. They have to be controlled by people exercising responsible human judgement. All people have to claim responsibility, not just those who initiate or develop a new technology.

In the example of word literacy, I see the main problem as arising from the way in which too many people accept a direct equivalence between knowledge and written expressions of knowledge, so that written expressions are used as though they were knowledge. This confusion is not inherent to our system for constructing written expressions. It arises from our use of this system, and originates from outside the system. Too many people attribute unnecessary (and magical) properties to written expressions. All expressions ought to be understood as representing certain people's knowledge, and they should be accepted as questionable by any other people. Writing ought to be accepted simply as a powerful mode of expression by which people can exhibit their responses to each other.

COMPUTER LITERACY

Computers offer a means for constructing and operating on expressions. In this respect, computer technology can be viewed as an extension of word literacy, offering systems for producing and modifying arrangements of written words. In Chapter 7 I indicated how such use of computers can be extended to include expressions in the form of drawings, and how a computer can be made to operate on a uniform representation of text and drawings. Now I will consider some implications of this kind of development.

Following on my discussion of word literacy, I want to consider whether computers can be used to alleviate some of the problems outlined, or whether computer technology is likely to lead us into even greater problems.

As in the case of word literacy, computer literacy refers to the structures of forms of expressions and our ability to use those structures to produce expressions of human knowledge. We are primarily interested in the structures which condition the production of expressions, independently of meanings that people might attribute to particular expressions. The major difference offered by computers is that they can serve as an active medium for expressions, in place of marks on paper. They offer an active medium which can execute actions on expressions.

Consistent with the strategy for CAD set out in Chapter 7, the actions that can be performed by a computer ought to be regarded as directly associated with the general structure for a form of expression. They should not be regarded as actions corresponding to behaviour of things that people might interpret from particular expressions. As said before, computers cannot know what such things are, in the sense that people can know them, and they cannot exercise responsibility for the behaviour of those things. Computer literacy, then, refers to people's ability to control the actions of computers on expressions, to produce expressions which reflect people's own knowledge.

Again, as indicated in Chapter 7, the effects of this position cannot easily be determined. By calling on the actions of a computer to modify an expression it is possible that the computer will exhibit behaviour that looks as though it is performing a meaningful task in terms of a person's interpretation of the changing expression. The computer might look as though it knows what it is doing. However, any such behaviour needs to be understood as a straight reflection of the person's intelligent use of the machine.

This point may appear a bit obscure, but it is vitally important to the concept of computer literacy. It is important to avoid the confusion that computers can know things like people can know things, which is likely to lead to the mistaken conclusion that knowledge in a computer can be the same as human knowledge. We also want to avoid the position in which knowledge represented in a computer prescribes knowledge in people. We do not want domain knowledge (meaning knowledge relevant to an application field, for which a person is responsible) to be represented in a computer in a manner that is invisible and that cannot be modified by the user of the computer.

If we fail to meet these conditions, then we will be faced with a dramatic worsening of the problems outlined previously under word literacy. The challenge of computer literacy is to diminish or remove those problems.

Computer operations
When computers are used as word-processors or drawing machines, they perform operations on their representations of objects and relations that constitute structures of expressions in the form of text or drawings. These structures are usually variants of the kinds indicated in Figures 8.1 and 7.7, which correspond to the ways in which expressions may be seen by users. Variants of these structures differ in their decompositions of text or drawings, and in the ways that they treat compositions. These differences condition how systems can access what parts of expressions, and what editing actions they can apply to those parts.

These systems are defined in terms of computer operations that can be applied to logical relations for objects that can occur in text or drawings. The effects of computer operations are seen as edits to states of text or drawings displayed on a screen. Users assess what they see, calling on any of their own knowledge outside the system which they think is relevant to the displayed expression. Users can then invoke further operations within the computer to effect further edits to a current expression.

To progress beyond currently established word-processors and drawing systems, we can define further logical relations of the kind indicated in Figure 7.6. Objects can be known to the system as being 'part of' other objects, and they can 'inherit' some of their parts from yet other objects. Parts of drawings can then become parts of text and vice versa, as constituents of a composite representation structure for this combined form of expression. Users can then use these objects and relations to model their own knowledge in a computer. The role of the computer system is to hold representations of users' knowledge in some coherent way by means of its logical relations applied to user declared objects, and the system displays any current state of its representation as expressions in a form familiar and visible to users.

Computer operations can be applied to any parts of its current representation of a user's knowledge, and an operation can be subjected to conditions that have to be satisfied in the representation before it can be executed. By editing expressions, the user invokes computer operations which change its current state of a representation. The role of the computer system then is to re-establish the coherence of its representation, resolving knock-on effects resulting from user edits to objects and relations. Just as with established word-processors and drawing systems, the effects of the computer operations offered by these more advanced systems are seen as computer generated edits to composite states of text and drawings on a screen.

Using computers
By using a computer in this way, it can be made to execute tasks that are meaningful with respect to a user's domain knowledge. Such machine behaviour can be useful in helping a user to explore consequences of proposed actions in the user's world, as in the case of designers exploring their own proposals for design objects,

To meet the conditions for computer literacy noted earlier, we have to hold on to the idea that any apparently meaningful behaviour of a computer remains a straight reflection of a person's intelligent use of the machine. Here we have the same issue that applies to word literacy: any impressive production of written words ought to be regarded as a straight reflection of a person's intelligent use of words. In both contexts, the artefacts serve as a means of interaction between people, allowing people to exercise and exhibit responsibility.

This ambition for computer literacy has major implications. It means that people who use computers have to know the logical relations and computer operations that can be applied to constituents of the formal structure of their expressions. They have to know how all this behaves in any instance of expressions, in order to use a computer to produce and modify expressions of their own knowledge. They must know all this well enough to be able to use computers as freely as we presently use the system of writing to produce written expressions.

There is no easy and quick way of getting people, in general, to be computer literate. What has to be known can be acquired only through gradual absorption of experience over many decades. Here we are anticipating an evolution of human knowledge which will have deep and unforeseen effects on how we view and use our world.

Progress
Computer literacy has been described as a long-term ambition. Meanwhile, using the parallel of word literacy, we are at the stage of development where we have to rely on scribes to program computers to do things for us. When we change our minds about what we want to do, we have to call in the scribes again.

Present advances in computer technology can be judged on two criteria. First, advances can be assessed in terms of how much they enable people to do with their computers, before they have to call upon scribes to program their computers — presently, Apple Macintosh computers offer a good example. Secondly, given that users will want their computers to do complex things in ways particular to their own needs, advances can be assessed in terms of how easily and quickly scribes can program and reprogram their computers — presently, Sun/Unix computers offer a good example.

Such technological advances will never be enough. They can even be counter productive if they mask what users need to know about a machine's logic for purposes of producing their own expressions. The more important advances will come from people's general and gradually increasing familiarity with the kind of logic that can be operated within computers, and their frustrations with current implementations. Progress will depend on system designers accepting and responding to criticisms coming from users' experience. We probably do not know yet what logic system can eventually support widespread computer literacy.

OUR FUTURE

Computer literacy, as outlined in the previous pages, will take a very long time to evolve and it might never be fully realised. Even if this notion of computer literacy eventually proves to be unattainable, it remains an important goal. This goal will prompt the kind of developments needed to alleviate the problems described in the middle chapters of this book — the problems that have become evident through users' experience of CAD systems. In my concluding paragraphs, I will assume that computer literacy is feasible and I will offer some optimistic speculation about its wider effects.

A computer literate society
Consider a future in which computer technology is accessible and usable by all people. People will then use computers to produce and modify expressions of their own knowledge. They will be able to store and recall these expressions for reuse as their own knowledge changes, and they will be able to access and assimilate expressions of each other's knowledge. People will use computers to exhibit their own intelligent behaviour among other people, and their expressions will exhibit their responsive behaviour to expressions of intelligent behaviour from other people.

People will be able to do this by employing the logic system which maintains the coherence of expressions and conditions operations on expressions, within computers. This system will be generally accepted among people and it will support freedom of expression without responsibility for interpretation within people. People will be able to use it to produce any expressions of any knowledge. They will be able to use the logic system to assimilate very many expressions from different people into coherent expressions which reflect a lot of human knowledge.

The criteria for correctness or acceptability of expressions will be the responses which they evoke within people. Knowledge will be recognised as being within people, and computers will be recognised as a means for expressing knowledge and passing expressions between people. Acceptance of computer-produced expressions will still be subject to more direct modes of expression between people — we will still be able to smile and laugh.

This general position is similar to our present situation which has resulted from our use of word literacy. There is, however, one important qualification and one important difference. The formal structure of written expressions (Figure 8.1) has been evolved by people and now we are anticipating some further evolution of such structures implemented within computers. If that happens, then the future logic system for maintaining coherence of expressions will need to be capable of accommodating yet further evolution — this qualification might be met, but we do not yet know how.

The important difference is that the processing of expressions, for purposes of executing mappings across expressions to reflect different states of knowledge, can employ the functionality of computers. Information will be less dependent on human processing, less labour intensive. People will be able to process large amounts of information without requiring other people to serve as bureaucratic information-processing machines.

Social structure and authority
By reducing the need for vast and costly armies of people to process information, computer literacy might have the effect of removing a primary justification for the existence of large centralised bureaucracies. By undermining the existence of such bureaucracies, computers would have the effect of diminishing their centralised authority. Instead, authority and associated powers of decision affecting the circumstances and actions of people might be claimed by those people, locally and individually. To see this as an attractive prospect, we need faith in people's recognition of their interdependent existence, and in their exercise of responsibility. We have here the recipe for democracy.

Access to information, in the form of expressions, plus the ability to assemble and recompose lots of information freely and easily will have profound effects on people's *beliefs*. More people articulating more information, with equal authority and in support of more divergent conclusions, could change the popular perception of knowledge. *Objectivity*, based on consensus of experience within specialist groups of people, may have to become pliable, yielding to knowledge within persons. We may see a shift in favour of greater recognition of contributions coming from people's *intuitions*.

Given such a trend, we will be faced with deeper changes. We may have to change the fundamental beliefs that underlie our perceptions of existence and our abilities to do things. We should expect to see evidence of such changes in the field of physics and in a blurring of the distinctions between 'hard' sciences, social sciences, and the arts. We may have to change the *value* that we attribute to the *individuality* of persons, to their holistic powers of judgement, and to their actions in any *community* of people. We should expect to see such change in a new recognition of human individuals as beings that are different from bureaucratic or artefactual machines. We should then expect to see the effects of these changes reaching up into the structures of our social and political institutions, resulting in a changed world of people. Such a transition would be strongly resisted by our present political, commercial, and academic authorities, but it might just possibly prove inevitable in human evolution.

Computer Discipline and Design Practice 229

Opportunities for designers
Emphasis on access to information, any information, and the ability to do anything with information is consistent with my earlier discussions on design. Designers need to exercise their intuitions on overt expressions of knowledge in response to any demands made on them by other people. Those demands typically are loosely defined and do not have calculable solutions. Designers have to exercise idiosyncratic design procedures by which they can synthesise design proposals, and they have to exhibit their proposals to other people. Designers have to be free to do so, unconstrained by prescribed problem decompositions and analytical procedures.

Growth in the use of computers as a means for realising expressions of knowledge now also impinges on the work-practices of designers. Designers are expected to use computers to exhibit and justify their design knowledge to other people. Designers, therefore, have an interest in becoming familiar with, and influencing, new advances in computer technology. Designers need a technology that can accommodate their holistic responses to human needs. This is a general requirement which applies also to people engaged in other fields — in art and science, and in the practical world of industry.

In the world of people, there is no reason to expect that the need for design will diminish. On the contrary, this need will increase as developments in science expose more uncertainty and if technology provides uncomfortable answers. Designers have to make this need evident by advocating the contribution which design can make to the well-being of people — they have to do so while also producing good exemplars of design. Designers have to enter into a wide debating arena, taking part in all issues that concern people. In order to hold their place in this arena, designers will need to become computer literate. Designers need access to information that will support their general advocacy of design, to win back individual responsibility for design objects.

If computer literacy can develop in the way I have outlined, it can have the effect of increasing the solution space for design objects. Serving as sensitive and responsive beings, exercising their trained experience on readily available information, designers should regain increased authority for design decisions. We should then expect greater empathy between the resulting design objects and the people for whom these objects are designed.

SUMMARY
This chapter has assembled a broad view of computer technology,

by referring to technology in general, the development of word literacy as a familiar technology, and the development of computers as a natural extension of that technology. All these developments need to be viewed as part of human evolution. Evolution here is focused on externalisations of knowledge and the need for reconciliation with people's intuitions. By presenting this view of computers I do not mean to imply that computers and the things we can make them do are, in some sense, inherently good. Rather, I mean to stress that our use of computers now is a part of our existence and that our future is inexorably bound up in their further development.

Computers as we now know them, and their current applications, do not define future computers. We have to decide what we want computers to be, while striving to realise our intentions, and we should expect to reshape our definitions. My purpose has been to show how we may come to change our perceptions of information and all the abstract and concrete mechanisms we employ to externalise human knowledge. Such change may affect our belief in the authority of overt expressions of knowledge, with further effects reaching into the practical, social, and aesthetic lives of all people.

So far, computers have benefited some people and have appeared threatening to many other people. They have hardened the separation between objective knowledge and intuition, obscuring access to information and dividing people. They are seen as imposing a strange and rigid formality which threatens to inhibit further evolution of knowledge. We have to believe that we can reverse these effects. Our future, including our future use of computers, depends on contributions from everyone. We need responsible and sensitive contributions from technologists and, more importantly, we want informed critical and creative contributions from users — scientists, artists, and designers, and people doing practical jobs in industry. We want criticisms to take us beyond present enthusiasms for current technology, to ensure a vigorous future.

BIBLIOGRAPHY

Alvey Committee (1982) *A Programme for Advanced Information Technology*, HMSO.

Bensasson, S. (1978) *Computer Programs for Continuous Beams—CP110*, Design Office Consortium (now CICA) Evaluation Report No. 2.

Bernstein, J. (1973) *Einstein*, Fontana.

Bijl, A. (1968) *Decisions Affecting Design of Five Point Blocks*, Ministry of Public Building and Works (now DoE) Report.

—— (1978) 'Machine Discipline and Design Practice', in proc. *CAD78*, Brighton, IPC, pp. 17–25.

—— (1979a) 'Computer Aided Housing and Site Layout Design', in proc. *PARC79*, Berlin, AMK/Online, pp. 283–292.

—— (1979b) 'Qualitative and Quantitative Functions within Computer Aided Design Systems', in Dierks, K. (ed) *Data Processing in Architecture*, Werner-Verlag, Dusseldorf, pp. 16–30.

—— (1983) *Architects and Computers: A Human Approach*, presentation to the Swedish Institute of Architects, SAR Meeting on Computers and Architecture, Stockholm.

Bijl, A., Nash, J. and Rosenthal, D. S. H. (1980) *Computing Practice Illustrated by Experience of a Ground Modelling Program*, EdCAAD wkg. paper, University of Edinburgh.

Bijl, A., Renshaw, T. and Barnard, D. F. (1970) 'The Use of Graphics in the Development of Computer Aided Design for Two Storey Houses', in Kusuka, T. (ed) *Use of Computers for Environmental Engineering Related to Buildings*, US Nat. Bur. of Stds., pp. 21–36.

Bijl, A. and Shawcross, G. (1975) 'Housing Site Layout System', *Computer Aided Design*, 7(1), pp. 2–10.

Bijl, A., Stone, D. and Rosenthal, D. S. H. (1979) *Integrated CAAD Systems*, EdCAAD Report for the Department of the Environment, University of Edinburgh.

Bijl, A. and Szalapaj, P. J. (1984) 'Saying What You Want with Words and Pictures', in proc. *INTERACT '84*, London, pp.(1)371–375.

Bronowski, J. (1971) *The Identity of Man* (revised edition), US Natural History Press.

Capra, F. (1983) *The Tao of Physics*, Fontana/Flamingo.

Clocksin, W. and Mellish, C. (1981) *Programming in Prolog*, Springer-Verlag, Germany.

Date, C. J. (1975) *Introduction to Database Systems*, Addison-Wesley, US.

Dawkins, R. (1978) *The Selfish Gene*, Paladin Granada.

Dreyfus, H. L. (1979) *What Computers Can't Do — The Limits of Artificial Intelligence* (revised edition), Harper Colophon Books, NY.

Eastman, C. M. (ed) (1975) *Spatial Synthesis in Computer-Aided Building Design*, Applied Science.

Flew, A. (1984) *A Dictionary of Philosophy*, Pan Books.

Frost, R. A. (1986) *Introduction to Knowledge Base Systems*, Collins.

Fuchi, K. (1981) 'Aiming for Knowledge Information Processing Systems', in proc. *Fifth Generation Computer Systems*, Tokyo.

Gifford, J., McWilliam, C. and Walker, D. (1984) *Edinburgh*, The Buildings of Scotland, Penguin.

Holmes, C. (1978) 'Plastic Merge Procedure for Mosaics of Polygons', *Computer Aided Design*, 10 (1), pp. 57–64.

Hoskins, E. M. (1977) 'The OXSYS System', in Gero, J. S. (ed) *Computer Applications in Architecture*, Applied Science, pp. 343–391.

JIPDEC (1981) *Preliminary Report on Study and Research on Fifth-Generation Computers 1979–80*, Japan.

Johnson-Laird, P. N. (1983) *Mental Models — Towards a Cognitive Science of Language, Inference and Consciousness*, Cambridge University Press.

Klein, E. (1987) 'Dialogues with Language, Graphics and Logic', in *ESPRIT '87 Achievements and Impacts*, North-Holland, pp. 867–873.

Kowalski, R. (1979) *Logic for Problem Solving*, North-Holland.

Krishnamurti, R. (1986) 'The MOLE Picture Book: On a Logic for Design', *Design Computing*, (1)3, pp. 171–188.

Krishnamurti, R. and Sykes, P. (1986) 'A Graphics Interface to Prolog', in Katz, P. (ed) *ESPRIT '85*, North-Holland.

Lansdown, J. (1982) *Expert Systems: Their Impact on the Construction Industry*, RIBA Report.

Magee, B. (1973) *Popper*, Fontana.

—— (1987) *The Great Philosophers: An Introduction to Western Philosophy*, BBC Books.

Maver, T. W. (1977) 'Building Appraisal', in Gero, J. S. (ed) *Computer Applications in Architecture*, Applied Science, pp. 63–94.

Minsk, M. (1975) 'A Framework for Representing Knowledge', in Winston, P. H. (ed) *Psychology of Computer Vision*, MacGraw-Hill, NY, pp. 211–277.

Mitchell, W. J. (1977) *Computer Aided Architectural Design*, Von Nostrand Reinhold, NY.

Michie, D. (ed) (1979) *Expert Systems in the Micro-Electronic Age*, Edinburgh University Press.

Negroponte, N. (1970) *The Architecture Machine*, MIT Press, US.

Norberg-Schulz, C. (1980) *Meaning in Western Architecture* (revised edition), Cassell Studio Vista.

Pears, D. (1985) *Wittgenstein* (with postscript), Fontana.

Bibliography

Pineda, L. A., Klein, E. and Lee, J. (1988) 'GRAFLOG': Understanding Drawings through Natural Language', *Computer Graphics Forum*, 7 (2), North-Holland, pp. 97–103.

Popper, K. R. (1963) *Conjectures and Refutations: the Growth of Scientific Knowledge*, Routledge & Kegan Paul, US.

Putnam, H. (1978) *Meaning and the Moral Sciences*, Routledge & Kegan Paul, US.

Pye, D. (1978) *The Nature and Art of Workmanship*, Cambridge University Press.

Russell, B. and Whitehead, A. N. (1910) *Principia Mathematica*, Cambridge.

Schumacher, E. F. (1974) *Small is Beautiful: A Study of Economics as if People Mattered*, ABACUS, Sphere Books.

Searl, J. (1984) *Minds, Brains and Science*, 1984 Reith Lectures, BBC Publication.

Sloman, A. (1978) *The Computer Resolution in Philosophy: Philosophy, Science and Models of Mind*, Harvester Press.

Steiner, G. (1978) *Heidegger*, Fontana.

Stiny, G. (1980) 'Introduction to Shape Grammars', *Environment and Planning B*, 7, pp. 343–351.

Sutherland, I. E. (1963) 'Sketchpad — A Man-Machine Graphical Communication System', *Spring Joint Computer Conference*, Spartan Books.

Szalapaj, P. J. (1988) *Logical Graphics: Logical Representation of Drawings to Effect Graphical Transformation*, PhD Thesis, University of Edinburgh.

Szalapaj, P. J. and Bijl, A. (1985) 'Knowing Where to Draw the Line', in Gero, J. S. (ed) *Knowledge Engineering in Computer-Aided Design*, North-Holland, pp. 147–164.

Tweed, C. and Bijl, A. (1988) 'MOLE: A Reasonable Logic for Design?', in ten Hagen, P. J. W., Tomiyama, T. and Akman, V. (eds) *Intelligent CAD Systems II: Implementation Issues*, Springer-Verlag.

Weizenbaum, J. (1976) *Computer Power and Human Reason*, Freeman, US.

Winograd, T. (1980) 'What Does It Mean to Understand Language', *Cognitive Science* 4, pp. 209–241.

Winograd, T. and Flores, F. (1986) *Understanding Computers and Cognition: A New Foundation for Design*, Ablex Publishing Corporation, New Jersey.

Woods, A. W. (1975) 'What's in a Link: Foundations for Semantic Networks', in Bobrow, D. G. and Collins, A. M. (eds) *Representations and Understanding*, Studies in Cognitive Science, NY Academic Press, pp. 35–82.

INDEX

abstractions, 29, 31, 43–4, 49, 53, 183
accountability, 61–2, 80, 104–5, 134
accuracy, 102, 103, 120, 126, 161, 167;
  see also correctness
aesthetics, 24, 33, 34, 55, 72–9
AI *see* artificial intelligence
alienation, 79, 214
Alvey Committee, 130n
animals, 32n
answerability, 61–2, 80, 104–5, 134
appraisal *see* evaluation
architects, 58–64, 79–80, 142, 143–7
  and CAD, 81, 134, 135, 158
  and function-orientated design systems, 120
  drawing practices, 152–4, 158, 170, 171
  using litigation, 222
  see also designers
architecture, 24, 33, 142
  and innovation, 62–4
  effects of mechanisation, 72–9
  examples, 13–17
  see also buildings *and* design
artificial intelligence, 4–5, 11, 48, 71–2, 128, 130n, 131, 179, 213–4; *see also* expert systems
assumptions, 22, 134; *see also* prescriptiveness

belief, 4
Bensasson, S., 132n
Bernstein, J., 63n
Bijl, A., 8n, 9n, 83n, 96n, 119n, 190n, 191n;

experience, 2, 13–21
Bronowski, J., 3n
building regulations, 133, 221
buildings, 24, 53–8, 63, 152–4
  and correctness, 69
  and logical expressions, 37–9
  costs, 55, 98, 103; *see also* quantity surveying
  decomposition of design, 67–8, 83, 94, 117–19
  responses to, 72–9
  see also architecture
bureaucracy, 221–2, 228

CAD, 25–7, 53, 81–2
  architects and, 81
  finances of, 97
  history, 82, 83
  limitations, 69, 70–1
  limitations on progress, 1–2, 8–10, 11
  strategy for, 12–13, 69–72, 80, 173, 175, 179–90, 207–9, 223
  see also design systems *and* drawing systems
Capra, F., 34n
change, 70, 175, 182, 191, 212–13;
  see also editing
Clocksin, W., 193n
common sense, 42, 43, 135
communication, 27, 28–30, 178–9
  and computers, 39, 131–2, 197–9, 207
  and descriptions, 175–6
  and drawings, 150
  see also language
component libraries *see under* libraries

235

Index

components, as units in integrated design systems, 88, 89, 90–2
computer literacy, 13, 113, 207, 212, 216, 223–9
computers, 6
  and design, 64 (*see also* CAD)
  and expressions, 39
  and knowledge *see under* knowledge
  and language, 23–4, 28, 39, 131–2, 135
  and limits of tasks, 70, 101, 102–3, 113, 119
  and logic *see under* logic
  and power *see under* power
  and representations, 43
  and responsibility, 45, 103
  and understanding, 36
  display screens, 161, 163–4, 167
  effects, 3
  generality, 43, 45
  intelligence and, 49–50
  'intelligence' *see under* machines
  successful applications, 45, 134, 135, 137, 224–5
  types, 97, 169–70, 226
concepts *see* expressions
contours *see* ground modelling
control, 115; *see also* prescriptiveness
correctness, 69, 126, 133, 143, 161
  in programming, 113–14
  see also accuracy

data, hierarchical arrangement of, 109–10
databases, 107
Date, C. J., 107n
Dawkins, R., 29n
decomposition
  of building design, 67–8, 83, 94, 117–19
  of design, 67–8, 83, 94, 117–19
  of drawings, 165, 167–8, 186, 188, 197
  of expressions, 178–9, 180
democracy, 222, 228

description systems, 11–13, 176–8, 191–2, 208–9
descriptions, 11, 175–8, 206; *see also* expressions
design, 3, 5, 6–8, 21–2, 24–7, 64–9, 80, 129–30, 209
  and language, 24
  and logic, 199
  decomposition of *see under* decomposition
  demarcation in, 70
  drawing in, 142–7
  finances of, 97, 98
  indivisibility, 67–8
  innovation and, 62–3, 64–5
  quality, 72, 98
  technology and, 215–16
  see also architecture, CAD *and* drawing design systems, 8–11, 53, 65, 67, 81–2
  function-orientated, 9–10, 116–35, 172
  integrated, 8–9, 81–115, 161, 172, 190–1
  users' experience of, 94–115, 120–8
  see also drawing systems
designers, 53, 70, 71, 175, 180, 182, 191, 229
  requirement to use computers, 216, 229
drawing, 140, 142–9, 170–1
  primitives, 148, 161, 164, 165, 186–90
  technology, 14, 147–9 (*see also* drawing systems)
  see also drawings
drawing machines, 194–5, 197
drawing systems, 10–11, 134, 135, 136–8, 154–71, 172, 224–5
  choosing, 169–70
  costs, 10, 169
  definitions and, 35–6
  users' experience of, 161, 164–5, 166–9, 171
  see also CAD (strategy for), description systems

Index

and logic modelling systems
drawings, 138–54, 170–1, 182–90,
  208–9
decomposing *see under*
  decomposition
editing *see* editing
in logic modelling systems, 194–5
see also drawing *and* pictures
Dreyfus, H. L., 3n, 48n

Eastman, C. M., 117n
EdCAAD, 119n, 190n, 191n
editing, 149, 157–8, 160–1, 162,
  166, 167–8, 177, 186–7, 188,
  189–90, 191, 224–5
Einstein, A., 34n, 63
equivalence, 32
evaluation, 9–10, 117–19, 126
in logic modelling systems, 205–6
evolution, 212–13, 216, 226, 228–9,
  230; *see also* change
expert modules, 180, 185–6, 188
expert systems, 11, 128–32, 135
experts, 9–10, 129
  designers as, 130
expression(s), 28–9, 31, 34–9,
  43–4, 49, 53, 72–9, 80, 135,
  143, 173–4, 176, 178–9, 183
  and CAD, 179–80, 183
  and computer literacy, 225–6,
    227
  and computers, 39, 50
  and logic modelling systems,
    194–9, 206, 208
  buildings as, 55
  graphical, 182–3
  in description systems, 191–2
  see also descriptions *and* words
externalisation 30, 33; *see also*
  under knowledge

Flew, A., 31n
Flores, F., 29n, 32n
FORTRAN, 107, 120, 128, 194
Frost, R. A., 107n
Fuchi, K., 130n
function-orientated design systems

see under design systems

generality, 43, 44, 154–7, 168–9,
  170–1, 180, 186, 188; *see also*
  prescriptiveness
geometry, limited, in integrated
  design systems, 83–4, 88, 94
GKS, 188n
Gödel, K., 31n
graphics *see* drawings
ground modelling, 119–28, 132

Heidegger, M., 32n
Holmes, C., 96n
Hoskins, E. M., 83n
houses, 54–5, 58
housing, 55–8, 95, 96

individuality, 173, 178–9, 228
  and knowledge, 40, 42
individuals, 29–33, 32–3
  computers and, 39
  in logical statements, 36–9
information *see* knowledge
inheritance networks, 189, 191,
  196–7, 206, 225
innovation, 62–5
integrated design systems *see under*
  design systems
intelligence, 3, 4–5, 32n, 46–8, 49,
  176
  and computers, 49–50
  and description systems, 12
  language and, 27–30
  non-human, 6, 27, 49 (*see also*
    artificial intelligence)
intentionality, 128, 132, 135
interfaces, 108
interpretations, 28, 29, 48, 49, 131,
  176, 177, 178, 206
  of drawings, 148–50, 151–2, 154,
    161–2, 183–6
intersections *see* junctions
intuition, 42, 43, 212, 213, 228
  and design, 63, 64–5, 129–30
  and expert systems, 129
  and intentions, 128

Index

architects and, 63, 64
  necessity for allowing use of, 113
JIPDEC, 130n
Johnson-Laird, P. N., 32n, 34n
judgement(s), 5, 6–7, 44, 63
  value, 24, 55, 72–9
junctions, 88, 187, 199–205, 206–7
  between houses, 106
  in SSHA design system, 89–90,
    101–2, 106

Klein, E., 24n
knowledge, 4, 6–7, 22, 40–3,
    212–13
  accessibility, 228, 229
  and computers, 40, 42, 45, 48,
    128–32, 172–3, 174, 193–4,
    223–4, 227
  and descriptions, 175–8
  and intelligence, 47–8
  and interpretations of drawings,
    149–50
  and language, 23–4
  and logic, 177–8
  and logic modelling systems,
    206–7
  and models, 25, 127
  and technology, 212, 213–15, 222
  and writing, 2, 219–23
  evolution, 226
  externalisation, 4, 5, 6, 7, 40,
    42–3, 143
  in quantity surveying, 104–5, 106
  intuitive, 7
  specialisation *see* language,
    specialised
knowledge-based systems, 130; *see
  also* expert systems
Kowalski, R., 193n
Krishnamurti, R., 188n, 190n

language, 27–31, 32nn, 49, 132,
    178–9
  abstractions for, 35
  and common sense, 42
  and knowledge, 23–4

and logic, 33–4
computers and, 23–4, 28, 39,
    131–2, 135
drawings and, 188–9
graphics as, 27
natural, 2, 27–30
prescriptiveness and, 174
specialised, 2, 13, 27, 220
see also words, written
language systems, 179
languages, programming *see under*
  programming
Lansdown, J., 128n
learning, 47–8, 219, 226
libraries
  component, 88, 94
  of construction details, 90, 94, 95,
    102
limits, defining, 70, 101, 102–3, 113,
    119, 128–30
lines, 35–6, 147–9, 161–2, 164,
    186–8
linguists, 24n, 30
literacy, 13, 212, 217, 218–23, 226,
    227
  computer *see* computer literacy
  see also knowledge
    (externalisation) *and* words
logic, 22, 27–8, 30, 33–4, 36–9,
    45–6, 49, 131
  and computers, 33, 34, 43, 46,
    128–9, 177–8
  and design, 175
  and expressions, 227
  and knowledge, 227
  and language, 44
  limitations of systems relying on,
    44, 46
logic environments, 131, 135
logic modelling systems, 190–208
logic programming, 193–4

machines
  information-processing, people
    as, 221–2, 228,
  'intelligence', 3, 4–5, 39–40,
    174–5, 176

Index

intelligent use of dumb, 5, 6, 13, 39–40, 175, 224
Magee, B., 3n, 43n
magic, 5, 220, 223
Maturana, 29n
Maver, T. W., 117n
meanings, 176, 179, 182, 206; *see also* interpretations
mechanisation, 72–9, 215
Mellish, C., 193n
memes, 29n
Michie, D., 128n
Minsky, M., 195n
Mitchell, W. J., 117n
models, 25, 136, 179
   drawings as, 149, 150–2, 161, 185–6
   see also ground modelling
MOLE, 190–208

Negroponte, N., 11n
Norberg-Schulz, C., 73n

objectivity, 31–3, 228
Oxford Region Health Authority, 88
OXSYS, 83–8, 94, 114

Pears, D., 32nn
philosophy, 3–4
   and artificial intelligence, 48
physics, 34
pictures, 28, 30, 31–2, 35–6; *see also* drawing systems *and* drawings
Pineda, L. A., 189n
points, 161, 163, 164–5
Popper, K. R., 43n
power, 213–14, 215, 221–2
   computer literacy and, 228–9
   computers and, 216, 229, 230
   writing and, 2, 219, 220–3
predictability, 213
prescriptiveness, 8–9, 11, 108–10, 112–14, 115, 131, 133–4, 135, 171, 173–5, 191, 192–3; *see also* generality;

see also dumbness *under* systems
printers, 164
problem solving, 64, 66–7, 69, 70–1
programming, 107–9, 110, 113–14, 120–6, 131, 135, 193–4
   and expert systems, 130–1
   correctness in, 113–14, 133
   for drawing systems, 166, 170
   languages, 44, 107, 193
   prescriptive, 108–10, 112–14, 115, 133–4
Prolog, 193–4
Putnam, H., 32n, 46n
Pye, D., 7n

quality, 72
quantification, 117–19: *see also* decomposition
quantity surveying/surveyors, 97, 98–107, 114

reality, 31–3, 49
reason/reasoning, 5, 49
   spatial, 182
   see also logic
representations, 43–4, 49, 173–4, 179–80
resolution, 164
responsibility, 44–5, 103, 113, 222
Russell, B., 42n

Schumacher, E. F., 3n, 7n
Scottish Special Housing Association *see* SSHA
Searl, J., 3n, 48n, 132n
semantic networks, 196n
senses, 39n, 42
shape grammars, 165n
Shawcross, G., 96n
similarity, 31, 69
simulations, 132
sites, 106
   integrated design systems and, 96–7
   see also ground modelling
Sketchpad, 1n

Index

Sloman, A., 48n
SMM (Standard Method of Measurement), 100
specialists *see* experts
SSHA, 94–107, 112, 114–15
  design system, 83–4, 89–94, 96–7, 100–2, 106, 107, 108–12, 114–15, 190–1
Steiner, G., 32n
Stiny, G., 165n
subjectivity, 31–3
Sutherland, I. E., 1n
Sykes, P., 188n
systems
  dumbness, 137–8; *see also* machines; (intelligent use of dumb)
  see also description systems, design systems, drawing systems, expert systems, language systems *and* logic modelling systems
Szalapaj, P. J., 188n, 190n

task-specific design systems *see* design systems, function-orientated
tasks, defining limits *see* limits, defining
three-dimensional projections, 136, 143, 161, 180
tree structures, 109, 193
truth, 33, 34, 177–8
Tweed, C., 190n

understanding, 36, 176
Unix, 120, 128, 170, 194

value judgements, 24, 55, 72–9
values, 182, 228

Weizenbaum, J., 79n
Whitehead, A. N., 42n
Winograd, T., 29n, 32n
Wittgenstein, L., 32n
Woods, A. W., 196n
word-processors, 134, 135, 224–5
  drawing systems and, 10, 136–8, 142, 160–1, 165, 168, 224–5
  logic modelling systems and, 194
words, written, 13, 35, 137, 138, 165, 176–7, 183, 217–21
  and authority, 2, 219, 220–3
  see also language, literacy *and* logic modelling systems